Progress in Social Psychiatry in Japan

Yoshibumi Nakane

Progress in Social Psychiatry in Japan

An Approach to Psychiatric Epidemiology

Springer

Yoshibumi Nakane
Department of Neuropsychiatry
Nagasaki University
Nagasaki
Japan

This English translation is based on the Japanese original SYAKAISEISHINIGAKU NO IMA © Yoshibumi Nakane, 2010 Originally published in Japan in 2010 by Nakayama Shoten Co., Ltd.

ISBN 978-4-431-54102-8 ISBN 978-4-431-54103-5 (eBook)
DOI 10.1007/978-4-431-54103-5
Springer Tokyo Heidelberg New York Dordrecht London

Library of Congress Control Number: 2012951285

Printed on acid-free paper

Springer is part of Springer Science+Business Media (www.springer.com)

Preface

When I became involved in several international joint studies related to psychiatry and psychiatric treatments, the experiences often made me think of psychiatry in Japan, which is the foundation of my study, and made me want, even in a small way, to show data unique to Japan and introduce them in other countries. At the same time, I found it difficult to establish a broad-based understanding of the term "EBM (evidence-based medicine)." I frequently quoted foreign data in research presentations and lectures, and I thought that the existing data unique to Japan were not exactly sufficient. As a result, I started specializing in social psychiatry and later in psychiatric epidemiology, and I have continued efforts to obtain specific psychiatric findings in the area where I live and about events that have occurred.

I gained the cooperation of more than 40 psychiatrists in Japan to compile data related to psychiatric treatments and psychiatry in this country and published them in 1969. Titled *Images in Psychiatry: Japan*, the book introduced the historical progress of psychiatry and the actual situation of various fields in Japan, including psychiatric health care and education, biological psychiatry, psychopathology, psychoanalysis, and transcultural psychiatry. Although books describing the Morita Therapy (a unique treatment form developed in Japan), anthrophobia (*Taijin-kyofu-sho*), and some other specific fields have been published in English to date, I believe my book was the first one to provide an extensive description of psychiatry in Japan. However, the English edition of the book was published by a French publisher, and I do not know how much impact the book had on research outside of Japan.

In addition, in 2011 I published the Japanese edition of *Progress in Social Psychiatry in Japan*, based on findings shared with colleagues mainly in my field of specialization. When I told my Western friends about the publication of the book, I was encouraged to publish it also in English. That is how this book came to be published. I will be very happy if this book can be of any help for the psychiatrists in the world to gain information and understand the situation in Japan.

Nagasaki, Japan Yoshibumi Nakane

Contents

Chapter 1
Introduction

To date, it has often been pointed out that understanding the biopsychosocial view is important in psychiatry. I have strived to carry out psychiatric treatment, education, and research respecting this view. Having said that, it is not easy to be familiar with all of the specialized areas involved, such as biology (physiology), psychology, and sociology. Over time, I chose the directions for my studies in psychiatric epidemiology and entered the field of social psychiatry.

Directly after graduating from Hitoyoshi Prefectural High School (Hitoyoshi city, Kumamoto prefecture), I entered the Faculty of Science at Kyushu University. When I became 21 years old in 1959, I transferred from the Department of Chemistry in the Faculty of Science to the specialized course for 4 years at Nagasaki University School of Medicine, although there was no special reason. When I started my student life in Nagasaki, I never thought that I would continue to live in Nagasaki for more than half a century since then. However, looking back on my past days, it seems that there were good reasons for living here after all.

Following my graduation from the School of Medicine, I completed a 1-year internship at the Japanese Red Cross Nagasaki Genbaku Hospital, thinking of becoming a psychiatrist. In the course of my graduate studies at Nagasaki University School of Medicine, as my research direction was gradually becoming established, I encountered several research topics which I could not accomplish anywhere else except Nagasaki. Each of those topics involved in an important area that cannot be ignored within the field of clinical psychiatry. For example, when I majored in psychiatry in the course of internal medicine as a graduate student, I was interested in the relationship between heredity and environment in relation to several particular forms of mental retardation. I have also focused on some geographical areas in Nagasaki and the nearby prefectures where mental retardation is common. Environmental factors as well as a high intermarriage rate are greatly correlated with the incidence of mental retardation. In order to determine biological findings, I often went to areas where mental retardation is common to ask not only the patients but also the relatives who have not developed the disease for cooperation in conducting various tests. Those studies included susceptibility to endogenous psychiatric

Y. Nakane, *Progress in Social Psychiatry in Japan: An Approach to Psychiatric Epidemiology*, DOI 10.1007/978-4-431-54103-5_1, © Springer Japan 2012

disorder in people expected to be genetically heterozygous, as well as phenylketonuria and inborn error of lipid metabolism.

Nagasaki is the only area that fulfills the conditions of the subject of the study of psychiatric epidemiology and social psychiatry that I am going to describe in this book. Through this study, my colleagues and I have made substantial progress in finding clues and footholds in understanding such phenomena. We express our appreciation to the people of Nagasaki. We could not achieve the study of psychiatric epidemiology and social psychiatry without the help of people in that city. I would like to explain their involvement in the following sections.

Furthermore, I would like to apologize in advance that honorific titles for personal names are omitted in the text.

Part I
Social Psychiatry and Psychiatric Epidemiology

Chapter 2
Definition and Historical Review of Social Psychiatry and Psychiatric Epidemiology

2.1 Social Psychiatry and the Japanese Society of Social Psychiatry

The term *social psychiatry* is now commonly used. The concept among researchers has been envisioned appropriately at the social level. Still, most hesitate to confirm a clear-cut definition of the field.

As described below, *social psychiatry*, edited by Katsumi Kakeda and Masaaki Kato (1970, Igaku-Shoin) [1], may be the only book with the term *social psychiatry* in the title published in Japan, other than *social psychiatry* released by the Japanese Society of Social Psychiatry (JSSP) in spring 2009 (Igaku-Shoin). Katsumi Kakeda also published the review titled *social psychiatry* in *abnormal psychology* (*E*) *social pathology*, the first part of the *abnormal psychology* course (Misuzu Shobo) [2] previously published in 1958. These works provided his comments on the concept and field of social psychiatry from a historical viewpoint, referring to its relationship with sociopsychological and political viewpoints; we can generally understand social psychiatry through his studies in these works.

In his description (1958), social psychiatry can be summarized as "a field involved in reflection on the social nature of various concepts in clinical psychiatry; the sociological and anthropological determination of various issues related to mental disorder and treatment of them as social phenomena; and the determination of various social pathological phenomena (including delinquency, crime, suicide, and prostitution) from the psychiatric perspective [2]." While he described in his 1970 book that social psychiatry includes social psychiatric approaches to social pathological phenomena and mental disorders, social psychiatry as part of industrial psychiatry, clinical psychiatry, and community psychiatry; and psychiatry as an ancillary science of social science and the humanities, he anticipated the establishment of social psychiatry as *a comprehensive human behavioral science* [1].

A keyword search on the Internet for *social psychiatry* brings up the JSSP first. The author of the present work, who has been a member of this society for many years and served as its president until recently, introduced the definition of social

Y. Nakane, *Progress in Social Psychiatry in Japan: An Approach to Psychiatric Epidemiology*, DOI 10.1007/978-4-431-54103-5_2, © Springer Japan 2012

psychiatry in JSSP. Social psychiatry is defined on the top page of this society's web site as follows: "social psychiatry is an academic field analyzing mechanisms of mental health issues comprehensively, including the biological, psychological, and social aspects, and using the findings to contribute to society [3]." Social psychiatry is considered to be a multidisciplinary field, based on psychiatry and connected with a wide variety of academic fields, including psychology, sociology, law, economics, and religious studies. Specific studies of social psychiatry are directed at (1) the relationship between mental health and society, (2) social characteristics of patients with mental disorders, (3) patients with mental disorders in communities, (4) a social scientific approach to psychiatric disorders, (5) a psychiatric approach to social pathological phenomena, including suicide, abuse, and crime, and (6) mental health activities.

In the current chapter, I will give a brief history of JSSP with the hope that it will be helpful to understanding social psychiatry.

The Japanese Community Psychiatric Association (JCPA), established in 1967, was considered to be the predecessor of JSSP. The Reischauer Incident,[1] which occurred 3 years before the establishment of JCPA, led to the amendment of the Mental Health Act of June 1965 for security reasons. In addition, at the 63rd annual meeting of the Japanese Society of Psychiatry and Neurology (JSPN; president Haruo Akimoto, director of the National Musashi Sanatorium and chairperson of the academy at that time, vice-president Akira Kasamatsu) was held in Tokyo in April 1966, "Community Psychiatry – Theory and Practice" was addressed as part of the symposia (sectional meetings). These events might have contributed to the establishment of JCPA in 1967. The session meetings of the 63rd annual meeting of the JSPN were printed in the *Psychiatria et Neurologia Japonica* at that time as follows [4]:

Theme of sectional meeting 4. "Community Psychiatry – Theory and Practice"
Session chair: Masaaki Kato (National Institute of Mental Health)
Subchair: Ichizo Sato (Choshi Municipal Hospital)

1. The position and role of psychiatric hospitals: Yasuo Okada (Matsuzawa Hospital)
2. Perspective on this symposium: Ichizo Sato (Choshi Municipal Hospital)
3. The realities of alcoholism in several regions: Hiroshi Mukasa, Takashi Ichihara, Akira Eto (Department of Neuropsychiatry, School of Medicine, Kurume University)
4. A study on toxoplasmosis in 3-year-old children, focusing on community issues: Tadashi Haneda, Junzaburo Shimizu, Shoji Yanagida, Hideo Yamasaki (Department of Neuropsychiatry, School of Medicine, Chiba University),

[1] In 1964, while serving as US Ambassador to Japan, Edwin O. Reischauer was stabbed by a mentally disturbed youth. He received a blood transfusion and recovered from the wound, but the transfusion inflicted him with hepatitis. He never fully recovered, and though he continued to work and lead an active life, he died of its complications after over 25 years. (Wikipedia contributors, "Edwin O. Reischauer," *Wikipedia, The Free Encyclopedia,* http://en.wikipedia.org/w/index.php?title=Edwin_O._Reischauer&oldid=412150287.)

Wataru Akamatsu Hiroshi Kumada (Department of Neuropsychiatry, Choshi Municipal Hospital)

5. Psychiatric problems in islanders: Tsunetoshi Hirayasu (Okinawa Prefectural Seiwa Hospital), Seijun Tatetsu, Gyo Toya, Motoaki Takagi, Masazumi Harada, Koichi Mimura (Department of Neuropsychiatry, School of Medicine, Kumamoto University)
6. The viewpoint and current status of family support groups of mentally ill patients—community-based measures: Fukuichi Furukawa, Takashi Suzuki, Kazuyuki Yano, Jitsuo Nagata (Ibaraki Prefectural Tomobe Hospital)
7. Psychiatric hospitals in the eyes of families of patients with mental disorders: Hirosaburo Ota (Ibaraki Prefectural Mental Health Center)
8. Various issues in regional mental health activities: Tsuyoshi Yamagoshi (Department of Neuropsychiatry, Faculty of Medicine, Gunma University), Kohei Minemura (Oshima Hospital), Masao Nakazawa (Department of Neuropsychiatry, Saku Central Hospital)
9. Social psychiatric studies in specific areas I—focusing on patients diagnosed with schizophrenia: Ryuro Takagi, Masaichi Kuniyoshi (Department of Neuropsychiatry, Faculty of Medicine, Kyoto University), Kiyoyuki Koike (Department of Neuropsychiatry, Faculty of Medicine, Kyoto University, Biwako Gakuen), Hitoshi Kawai, Akihiko Kimura, Isao Ozawa (Department of Neuropsychiatry, Faculty of Medicine, Kyoto University), Haruo Kuwabara (Department of Public Health, Faculty of Medicine, Kyoto University)
10. A survey on attitudes of patients with mental disorders in communities: Shogo Terashima (Fukuoka Prefectural Sanatorium Tsukushi), Seisuke Matsuo, Kenichi Mikajiri (Department of Neuropsychiatry, Faculty of Medicine, Kyushu University)
11. From the viewpoint of public health activities: Masami Hashimoto (Institute of Public Health)
12. Psychiatric treatment and public health: Hideyo Kosaka (Matsuzawa Hospital)
13. General physicians and community psychiatry: Takeshi Kawakami (Suginami Union Hospital)

Special lecture: A concept of community in Japan: Goro Hani

General overview: Masaaki Kato (National Center of Neurology and Psychiatry)

Both chairpersons of this symposium subsequently became the first and second presidents of JSSP, and other panelists held important roles at Gunma University, Kyoto University, Tokyo Metropolitan Matsuzawa Hospital, and Kyushu University at that time. Many active psychiatrists with a social psychiatric perspective participated in the meeting. I can imagine heated debates on mental health practice and skills for community care at the meeting. In addition, these titles of the session presentations indicate that the directions of the practices and studies continue to remain alive. I imagine that a large audience enjoyed the presentation after the meeting by Takeshi Kawakami (1925–2009), an author of *Nihon no isha* (physicians in Japan, 1961, Keiso Shobo), with a detailed knowledge of social medicine and patients' cases in Japan, and the special lecture by Goro Hani (1901–1983), a well-known historian and philosopher who published an excellent work on his theory of community, titled *The Logical Understanding of Cities* (1968, Keiso Shobo). However, having just graduated from the university in March 1963, I could not well understand the efforts of those already working in the field at that time.

As described above, the Reischauer Incident (March 24, 1964) triggered an amendment of the Mental Health Act in June 1965, mainly for security reasons. Psychiatric health-care professionals, patients, patients' families, and medical staff at that time must have felt a very strong intended direction for social defense and security in the amendment. The sectional meetings of the 63rd annual meeting of the JSSPN were held the following year, April 1966, with inner resistance to the amendment, and most of the panelists of the social psychiatry symposium joined the JCPA founders' list.

The first annual meeting of the JCPA was held in Gunma with about 300 participants. Subsequent annual meetings were held in Kyoto, Tokyo, Miyagi, and Nagano, and the number of participants rapidly increased. However, potential controversies over technology and principles of community psychiatry came to the surface rapidly as well. Importantly, dissension occurred, mainly in the JSPN academy, contributing to cessation of its activities after the 6th JCPA annual meeting (Hakone, Kanagawa) in 1972.

For reference, the Japanese Hospital and Psychiatry Forum, established in 1957, 10 years prior to the establishment of the JCPA, changed its name to the Japanese Hospital and Psychiatry Association in 1967 and then to the Japanese Hospital and Community Psychiatry Association after 1983, for the purpose of developing clinical practice in communities. However, the association says that it is not part of the aforementioned JCPA even though it contains *community* in its name.

Information about the JCPA can be reviewed in articles in journals and related magazines. To some extent, the association might vacillate in relation to various protest movements against authorities. There is no question that community psychiatry is an important part of social psychiatry; we appreciate the efforts of the seniors in the field and their intentions to activate and develop theories for psychiatric care and mental health activities outside of a hospital (or in communities). However, the social circumstances have changed markedly in more than 35 years, and the radical movement of that earlier time has receded. While our life has improved physically, it is also true that most people face difficulties living in a society highlighted by the principle of competition in economic and psychosocial terms.

The need to promote social psychiatry studies suitable for the times has increased in western countries, the World Health Organization (WHO), and across the world. In Japan, this led to the publication of the *Japanese Journal of Social Psychiatry* in 1978 (Seiwa Shoten, quarterly). This journal allows psychiatrists to publish the findings of their psychosocial and epidemiological studies and research studies based on practices of psychiatric care and to exchange information. The editorial committee at that time included Masaaki Kato (then executive director of the National Institute of Mental Health) as editor-in-chief and Tadao Miyamoto, Kenshiro Ohara, Ichizou Sato, and Yukio Ishihara as editorial board members.

The JSSP was established at the then-National Institute of Mental Health on April 11, 1981, represented by Masaaki Kato, for the purpose of promoting research in the field of social psychiatry and contributing to the field's progress, development, and dissemination. JSSP made a new start for the first time in about 10 years after the cessation of its activities, and the first JSSP annual meeting was held in a tense atmosphere due to a legacy of continuous dissension. After this long interval, nearly 200 participants gathered in the meeting room of the institute and held heated debate. At that time, my colleagues and I reported on an ongoing study on seasonal birth in patients with schizophrenia and on part of the WHO Collaborative Research programs on the effects of the first episode of schizophrenia on patients' life events.

After 1981 and for some time, JSSP held the first and some subsequent annual meetings in Tokyo and Osaka. The 8th annual meeting was held in Nagasaki in 1988 to explore the feasibility of local venues. The success of this meeting allowed subsequent annual meetings across the country.

The *Japanese Bulletin of Social Psychiatry* was first published in 1993, improving on *Japanese Journal of Social Psychiatry* (its final edition vol. 15-4 was published in December 1992, the 58th issue). The concept of social psychiatry was regularly discussed for some time starting with the bulletin's first issue, and the research field, objectives, and methodology were introduced. The journal has already gone through its 20th volume (the 53rd issue was as of summer 2011, and the articles published in it show the direction of trends in social psychiatry in Japan).

The recent 30th annual meeting of JSSP held in Nara in March 2011 was chaired by Nara Prefectural Medical University. When my staff at Nagasaki University and I chaired the 8th Annual Meeting in Nagasaki some 20 years ago, we asked Robert P. Liberman, professor of psychiatry at UCLA School of Medicine, to give a special lecture titled *Psychosocial Management of Schizophrenia: Overcoming Disability and Handicap*. He introduced the topic of care systems, including social skill training, in Japan for the first time, which was a very good opener for future implementation of such systems in Japan.

Today, the society has grown to have about 1,600 regular members. Previous presidents are Masaaki Kato, professor emeritus at Tokyo Medical University; Ichizo Sato, former professor at Chiba University; and Kenshiro Ohara, professor emeritus at Hamamatsu University School of Medicine. In addition, I and Tetsuya Hirose, professor emeritus at Teikyo University, held the position in April 2004 and April 2008, respectively. A total of 56 board members (including 1 chairperson and 10 executive board directors) and 2 inspectors run the society.

JSSP hosted the 18th World Congress of Social Psychiatry (Kobe) October 23–29, 2004, with the main theme being *Globalization and Diversity*. We asked Kenzaburo Oe, a Nobel laureate in literature, to give the special lecture. I recall that, in answer to his question about social psychiatry and request for some reference material, we provided the above-described explanation. He gave a lecture titled *Write in Defiance of Violence,* and we were able to conclude that international meeting successfully. We were able to publish the first English edition of the journal as a special issue of the regular journal (*Japanese Bulletin of Social Psychiatry, vol. 14 supplement, 2006*) and report the proceedings. In addition, the JSSP edited and published about 450 pages of *Social Psychiatry in spring 2009*; this is also sort of an answer to Kenzaburo Oe's question.

2.2 Psychiatric Epidemiology

Epidemiology has been previously defined as *a study of the distribution of disease in time and space and the effects of various factors on this distribution* [5] and *a study of disease frequency in humans and its determinants* [6]. The Council for International Organizations of Medical Sciences (CIOMS) defines it as *a study of the distribution of health-related conditions and events in specific populations and their determinants and the application of studies for controlling health problems* (1991) in its *International Ethical Guidelines for Epidemiological Studies* [7]. These definitions indicate that epidemiological research in psychiatry is a field of study in patients with psychiatric disorder attempting (1) to investigate causes, (2) to develop clinical applications of the results, and (3) to determine applications to health services plans for individuals and/or communities.

However, investigation of causes in psychiatry has more than a few difficult problems. Concomitant factors (e.g., biological factors, including the influences of physical disorders, and social factors, including social stratification), precipitating factors (e.g., stress events), and predisposing factors (e.g., genetic factors, embryonal and prenatal factors, and environmental factors) should be examined for both independent and possible interrelated effects. Although, presently, biological factors are considered to be possibly involved in psychiatric disorders, dynamic and comprehensive study of the involvement of the other above-mentioned biopsychosocial factors should be considered also.

I describe this field in detail in the following chapter.

References

1. Kakeda K, Kato M. Social psychiatry. Tokyo: Igaku-Shoin; 1970.
2. Kakeda K. Abnormal psychology. Tokyo: Misuzu Shobo; 1958.
3. http://www.jssp.info/. Accessed Oct 2009
4. Akimoto H, Kasamatsu A. Abstracts of the 63rd annual meeting of the Japanese society of psychiatry and neurology. Psychiatr Neurol Jpn. 1966;68:115–301.
5. Lilienfeld AM. Epidemiological methods and inferences in studies of noninfectious diseases. United States Health Sci Rep. 1957;72:51–60.
6. MacMahon B, Pugh TF. Epidemiology: principles and methods. Boston: Little, Brown; 1970.
7. Council for International Organizations of Medical Sciences. International ethical guidelines for epidemiological studies. Switzerland: Stationery Office Books; 1991.

Chapter 3
The Foundation of Psychiatric Epidemiology Research

As previously mentioned, psychiatric epidemiology is an interdisciplinary research and a practical academic field that developed basically through psychiatric and epidemiological approaches, with the cooperation of many other fields, including public health, mental health, and sociology. Therefore, I would like to make clear the following key issues to be considered before launching a psychiatric epidemiology study, which can also serve as the foundation of social psychiatry.

3.1 Methodological Foundation

There are several classifications of study methods in epidemiology [1]; the researcher will choose whether to employ an observational technique or an experimental technique, whether to design a cross-sectional survey or a longitudinal survey over the course of time, or whether study cases are to be collected through a retrospective or synchronic, prospective/anteretrospective study.

The first issue to be decided in psychiatric or mental health epidemiology research is which methods among the aforementioned classifications should be employed. Where a study is carried out in daily clinical practice, a cross-sectional study using an observational approach is the easiest. However, the findings of such studies are usually limited and difficult to generalize. Therefore, even though an epidemiological study should ideally be carried out in daily clinical practice, it would be advisable to take a concurrent or longitudinal approach according to a clearly defined plan. Yet, because a great deal of human and financial resources are required to carry out a study and there are possible time constraints, researchers may have no choice but not to launch the study or to give it up halfway through. Therefore, when a researcher decides to conduct a large-scale study, it is important to carry it out in a phased and well-planned manner. The researcher may first obtain sufficient background information using either a cross-sectional or retrospective study before moving on to a prospective, longitudinal study. Though retrospective surveys tend to be undervalued, we cannot underestimate the importance of this type of study in finding

Y. Nakane, *Progress in Social Psychiatry in Japan: An Approach to Psychiatric Epidemiology*, DOI 10.1007/978-4-431-54103-5_3, © Springer Japan 2012

extremely rare events at the time of the study as well as the utility of such a study as a preliminary survey for an extensive epidemiological research project. Retrospective approaches might be even more highly regarded rather than a prospective study that has poorly designed survey items or has well-designed study questions with poor answers. In other words, the *quality of the data* is more important than the temporal direction of the study.

In actuality, experimental techniques are not to say exceptional but they are rarely used in epidemiological studies. They are most popularly used in certain areas such as determining the effects of various treatments, such as pharmacotherapy, or determining the effect and usefulness of psychosocial therapy. In these areas, prospective, longitudinal studies are commonly conducted by targeting cases within given criteria, making clear statistical analysis possible.

In addition to the choices of study methods, there are some underlying difficulties that are specific to psychiatry. The most significant issues stem from the psychiatric disorders themselves that are to be studied. First, there is the issue of the disease concept. In particular, this relates to the difficulty of the identification or determination of disorders, which occurs even where the definitions and scopes of the characteristic symptoms, and the definitions and perceptions of onset, progress, and outcomes have been widely agreed upon. Unless the identification criteria of the disorders to be studied are clearly defined to enable every psychiatrist in the study to follow a uniform guideline, it will be difficult to obtain comparable, highly reliable epidemiological findings. In the late 1960s, the US/UK project, a collaborative study focused on schizophrenia and mood disorders, driven by the high prevalence of schizophrenia in the USA, was launched by psychiatrists in New York in the USA and in London in the UK, both English-speaking countries. The study revealed that the concept of schizophrenia in the USA not only overlapped the concept in the UK but also included some parts of the definitions of depression, mania, personality disorders, and neurotic disorders. As a result, schizophrenia was diagnosed twice as often in the USA as in the UK. On the other hand, mood disorders such as depression were diagnosed in the UK three times as often as in the USA. These findings highlighted the differences in the disease conceptualizations or diagnostic preferences between the two countries (see Fig. 3.1) [2].

Similar comparative studies were actively conducted around world, triggering a surge of interest in the conceptualization and diagnostic methods of psychiatric disorders, further accelerating the move toward the development of diagnostic criteria. In particular, the diagnostic guidelines and criteria described in the Classification of Mental and Behavioral Disorders (ICD-10/F) chapter in the *International Classification of Diseases and Related Health Problems, Tenth Revision,* issued by the World Health Organization, and the *Diagnostic and Statistical Manual of Mental Disorders, Fourth Edition* (DSM-IV-TR), released by the American Psychiatric Association, are widely known and employed for clinical use and for researches, with several interview methods developed for each use. These guidelines and criteria allow us to have organized clinical data and enable us to obtain standardized, uniform study cases.

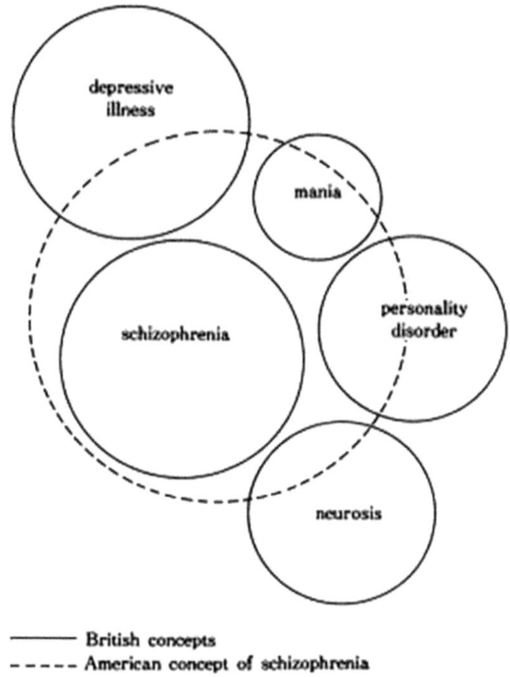

Fig. 3.1 Diagnostic preferences of schizophrenia in the US and UK projects [2]

The second research issue to be decided is the method of selection or sampling of the subjects. The selection criteria should specify the geographical areas to be covered as well as the characteristics of patients to be included in or excluded from the study. When a study focuses on the patients who are being treated at a particular medical institution, a researcher may be able to obtain an interesting outcome through highly accurate data analysis. However, the findings of such a study are specific to the particular institution and may be difficult to generalize. A long time ago, Masaaki Kato used the term *caseness* to highlight the importance of distinguishing between diagnosed cases and non-diagnosed cases [3].

In addition to the issue of caseness versus non-caseness, I would like to pay attention to *representativeness*. That is, it is important to know that the advantages in generally applying survey findings vary depending on what population the subjects represent. If researchers fail to select the subjects appropriately, all their efforts to attain results might go to waste. For instance, statistics about outpatients and inpatients, commonly reported from university hospitals or large-size hospitals, can vary depending on the degree of interest of the hospital staff, and the usefulness of the statistics can be clearly assessed only when such background is taken into account. If I look at a report from a university hospital on long-term outcomes of patients with epilepsy in which the seizure inhibition ratio is excessively low or the social functioning of the patients remains at an extremely low level, I would hardly say these are common outcomes in patients with epilepsy and would conclude that the results are instead specific to this particular hospital. That is to say, the results may have derived from the highly expected particularity that a university hospital specializes in extremely intractable cases.

The third research issue to be planned is assessment methods (measurement). Unfortunately, it is hard to deny that the reliability of the assessment of psychiatric symptoms has yet to be firmly established. Critics often point out the significance of the variances in clinical assessment and diagnosis of symptoms. In such circumstances, we have been expected to establish the reliability and validity of the assessment methods we use, and we are still far from attaining success. According to Spitzer et al. [4], variances are grouped into the following categories: subject variance related to the subjects participating in the study, occasion variance based on the survey period, criterion variance due to the diagnosis criterion employed, information variance due to the amount of information available for evaluation, and observation and interpretation variances due to the levels of observation skills being employed [4]. What we do today is to be well aware of these variances and try to improve each of them.

Various types of assessment methods and instruments have been developed in recent years. Usually, they are classified into three categories: structured self-report forms, symptom rating scales or symptom checklists, and structured or semi-structured interview formats that provide interview forms with specific procedures. While various types of structured self-report forms are available, with coverage ranging from general psychosomatic conditions to the conditions specific to each psychiatric syndrome, researchers should take into account the premise that these are self-assessments by the subjects. That means that the method is useful for identifying the complaint rate but is not necessarily directly applicable to clinical diagnosis. Furthermore, many structured self-report forms have been released in several different versions, ranging from complete to reduced versions, and researchers need to be aware of rating reliability or validity gaps among the versions. Therefore, assessment using a structured self-report form should be followed by a detailed symptom assessment or a psychiatric interview. Ultimately, psychiatrists need to carefully select the assessment instrument that suits the purpose of their studies. Assessment instruments developed overseas have been translated into Japanese for domestic use, which is obviously an advantage in terms of comparison among countries. However, some Japanese versions of assessment instruments have been intentionally modified to fit uses in Japan, and therefore the validity of the Japanese versions would also need to be verified through translation/back-translation under firmly established translation rights.

3.2 Ethics in Study Methods

Ethics in research, particularly those involving human subjects, has been emphasized often enough to make those who intend to pursue research think twice, even without the need to recall the declaration of Helsinki. Obviously, ethical considerations are an important issue, but this should not stop researchers from moving forward. To give up means admitting that the study is rather not worth doing in the first place.

In other words, if the researcher thinks the project is truly necessary, he or she should sincerely explain this, in a simple but thorough manner, to the subjects and tell them the study will be carried out based on ethical considerations, to persuade them to participate.

During the period when I worked as a permanent member of the institutional review board on the faculty of medicine at the university, I often came across studies whose authors seemed not to be aware of in which areas of their work ethics should be considered. In medicine or medical care, by rights, ethics in daily clinical practice should be the basis of ethical considerations in research. Therefore, the physician should secure ethics for human subjects in his or her clinical aspects before study-related ethics. Ethics may have been neglected where a doctor treat patients in a paternalistic manner in daily clinical practice. If ethics consideration is always appropriately practiced for patients or medical care users, it should not be so difficult to put such practices into a study.

Here I would like to stress again that ethics should be considered before beginning a research study, focusing on respect for the human rights of the subjects from the viewpoints of protecting personal information and respecting privacy. This means that the ethics of conducting a survey as well as of disclosing the outcomes of the study are in question here. The study should not only be valuable for the researcher but also be beneficial to the subject somehow. We should avoid studies of personal interest which are unlikely to attain any outcome to feed back to the subjects. When I have seen or heard about psychiatrists visiting the site of a disastrous event for their disaster psychiatry studies, their behaviors seemed like buzzards attacking their prey, and that made me question their honor as psychiatrists.

When researchers target human subjects, it is important that they provide the subjects with sufficient information to help them understand well the nature of the study and why it is necessary, so that they are able to determine whether to agree to cooperate with the research project autonomously. At the same time, it is a matter of course to seek a judgment and receive approval from the institutional review board at the target institution. Approval from the institutional review board is only a ticket to enter the study. There are still a few steps to take before moving on to the study. The subjects should be provided with sufficient information and written informed consents obtained from them. These steps are not to be missed before starting a survey. An outcome report should elaborate what information was given to the subjects and on what matters consent was obtained. The consent form should note that *prior to this study, approval was obtained from the institutional review board at the ABC Institution* and provide the subject with a detailed explanation about the study, with some reference information such as the approval number assigned by the institutional review board and the date of the approval, so that the series of steps taken is clear. As for the way to describe consent to participate in a study, reports from countries other than Japan usually state that *consent was given by the subjects,* whereas many reports by Japanese researchers in foreign languages state that *consent was obtained from the subjects.* This seems to reflect a difference in sensibility toward the autonomy of subjects.

What I describe here may sound pessimistic and discouraging for those who have intentions to pursue epidemiological studies. If a researcher plans a study project merely for personal interest, that researcher may fail to give a convincing explanation about the intended study and may end up with frustration. In the end, the extent of social needs that are met by the study is the key to determining whether the study is worth doing.

How we should provide feedback about the epidemiological study outcomes is another ethical issue. We might choose not to provide any feedback. However, preferably, outcomes should be disclosed to individual subjects upon the individuals' approval or request, and so should the overall findings, upon agreement by the subjects (although this should have already been approved by the institutional review board). Subjects' requests should not go unanswered even when significant findings were not obtained in the study. Since the study findings may be provided to the subjects and other stakeholders at a briefing session or by issuing comprehensive booklets, it is best to include the expenses incurred for such activities. Figure 3.2

Fig. 3.2 The booklet including the results of the mental health study on atomic bomb survivors

shows the booklet (approximately 40 pages, A4 size) of the outcomes of my mental health study on atomic bomb survivors, conducted over 2 years starting from October 1994, published to offer to the survey participants. Although the booklet was published by a department in the Nagasaki prefectural government, I was the one who compiled the articles on several mental problems explained in an easily comprehensible way by a few psychiatrists who joined my research study, as well as a summary of overall outcomes (published in March 1998). The chapter titles in the table of contents are as follows: (1) stress and stress release, (2) insomnia and alcohol, (3) depression, (4) anxiety/panic, (5) mental illness appearing in the body, (6) dementia, (7) the outcomes of the mental health study on A-Bomb Survivors at Nagasaki Heart Center.

3.3 Subjects of Epidemiological Studies in Psychiatry

The most typical factors in psychiatric epidemiological studies include prevalence rates, incidence rates, morbidity risks or lifetime expectancies, and mortality rates from various mental disorders, all which have been verified through carefully designed work across the world. In particular, Germany and the Scandinavian countries have been the leading countries in accumulating knowledge in the field. Also, in Japan, there have been nearly 30 *genetic epidemiological* field study reports based on traditional German or northern European methodology, including the study on Hachijo-jima Island (Tokyo) by a pioneering figure, Haruo Akimoto, and his colleagues at Tokyo University. In recent years, survey methods have been developing toward globally standardized methods utilizing rating scales based on the interview method, which enable researchers to collect detailed and uniform information and improve inter-study comparability.

As for the outcomes of these studies, many review reports have been published both in Japan and overseas, and information on outcome variability is gradually being clarified. During the period several decades ago in which specific outcomes were actively reported, outcome variability tended to be attributed to differences among the study methods. However, in recent years, this variability has been accepted as due to regional differences.

Professor Scott Henderson in Australia suggests in his work (*An Introduction to Social Psychiatry, 1988*) [5], based on his experiences, that epidemiological findings should include not only the incidence rates of disorders but also the elucidation of the variables in the etiology-related factors and mental-health-service-related factors, as shown in Table 3.1. He explains that, after the sociodemographic factors are identified, social data about the subjects' living environment should be extensively assessed, followed by the identification of the subjects' personal background factors. The variables include those which have already been identified as high relative risks/attributable risk factors in some psychiatric disorders and are also listed as the factors to be studied further. Premorbid personality and familial loading of psychiatric disorders are also critical survey items not to be overlooked. Furthermore, it is

Table 3.1 Variables constituting epidemiological findings

Etiological factors
• Sociodemographic factors Sex, age, marital status, social class, urbanization, regional variation, secular changes, and other macro-social factors • Experimental factors Seasonarity of birth, birth order, parental age, dissociation childhood separation, parental style, adverse experiences and events, extreme experience, bereavements, expressed emotional, unemployment, social support, migration, noise, environmental toxicity, and dietary factors • Personality and temperament • Familial history • Biological factors • Interactions of two or more factors • Natural history
Factors related to mental health services
• Social consequences on the family and the community • Services: proportion reaching primary care, proportion reaching psychiatric/specialist services, and social management or specific treatment • Knowledge available for: primary, secondary, and tertiary prevention • Evaluation of services • Tertiary prevention

better to look into the interactions between each listed factor as much as possible. At the same time, from the viewpoint of mental health services, several factors are provided in order to draw specific useful information and to feed it back to the subjects.

Other critical research issues include the appropriate assessment of how a subject and his or her family and local community have been affected by the psychiatric disorder, a determination of the ease of access to primary care, which is where the first action is taken, the possibility of approaching specialists as the next step, and an evaluation of the availability of social care and management. At the same time, an epidemiological study on the possibilities of reflecting on so-called preventive measures is another important issue.

Because of the trend in recent psychiatric studies, I believe that from now on the epidemiological factors shown in Table 3.2 need to be established. In other words, we will be requested in the future not only to pursue known epidemiological factors or etiological factors but also to set the assessment and treatment goals with consideration of the impact of the disorders on the social functioning of the affected person, as well as to establish epidemiological evidence that suits the goals set, while taking into consideration particularly the degree of satisfaction and the quality of life of the person. In order to do so, we will be expected soon to discuss the development of assessment scales suitable for all types of evidence and their availability.

Table 3.2 Epidemiological study fields

Fields	Typical result indices
• Frequency of psychiatric disorders	Incidence rate, prevalence rate, morbidity risk, mortality rate (suicide rate) and life expectancy, and the backgrounds of those factors
• Assessment of etiological association	
Biological factors	Genetic factors, data on patients' family and close relatives, susceptibility to physical disorders
Development factors	Embryonal life events, prenatal problems, development during childhood and adolescence, human relations during school age and young adulthood
Environmental factors	Economic factors, living environment, and legal problems, etc.
• Assessment of clinical characteristics	Expression frequencies of symptoms (syndrome) and problematic behaviors and the impact on social functions
• Frequencies related to physical complications and psychiatric complications	Types and frequencies of complications, interactions and impact of those complications with the primary disease
• Frequencies related to clinical course	Presence and level of symptomatic cycles, rapid remission, chronic course, and mental decay
• Assessment of treatment response	Drug selection criteria, response evaluation criteria, treatment algorithm, and side effect evaluation
• Assessment of treatment goal	QOL, satisfaction and social function levels
• Outcome assessment	Symptomatic outcome, socio-functional outcome, life prognosis, factors contributing to prognosis or outcome

References

1. Cooper B, Morgan HG. Epidemiologische Psychiatrie. Forschritte der Sozialpsychiatrie 3. Urban & Schwarzenberg, München, Deutschland; 1977 [Epidemiological psychiatry. CC Thomas Publ., Sprinfield, 1973].
2. Kendell RE, Cooper JE, Gourlay AJ, et al. Diagnostic criteria of American and British psychiatrists. Arch Gen Psychiatry. 1971;25:123–30.
3. Kato M. Problems of caseness and illness in multiaxial diagnoses in 9th ICD and DSM-III. Tokyo: Data on Mental Health 22, National Center of Neurology and Psychiatry, National Institute of Mental Health; 1976.
4. Spitzer RL, Endicott J, Robins E. Clinical criteria for psychiatric diagnosis and DSM-III. Am J Psychiatry. 1975;132:1187–92.
5. Henderson AS. An introduction to social psychiatry. Oxford: Oxford University Press; 1988.

Part II
Findings on Social Psychiatry and Psychiatric Epidemiology in Japan

Chapter 4
Activities of the Department of Neuropsychiatry Nagasaki University School of Medicine, as a World Health Organization Collaborating Center

Despite the increased importance of *evidence-based medicine* (referred to as EBM hereinafter) becoming known nationwide, there exists far less than sufficient evidence in Japan itself for field of psychiatry, not to mention the situation in other fields of medicine. I myself have been reiterating that efforts should be made to accommodate studies that aim to establish EBM. Referring to data published in foreign countries is considered a potential alternative; however, some data may show ethnic differences, and others may further suggest differences due to psychosocial and cultural variabilities.

To date, we have been involved in social psychiatric and psychiatric epidemiological studies in our own research studies conducted at Nagasaki University and those conducted by the designated WHO Collaborating Center for Research and Training for Mental Health almost concurrently. In addition to a part of our department's own research discussed in the previous chapters, neuropsychiatric epidemiology and social medicine research are also ongoing. Among these, details on radiation exposure of atomic-bomb survivors will be reported separately, not in this chapter.

As an example, my long-term effort has been personally dedicated to clinical research on epilepsy. In particular, great focus in epidemiological analysis was put on chemotherapy with new drugs developed for epilepsy treatment, review of effects and adverse effects of conventional antiepileptic drugs, clinical characteristics of intractable cases, and so on.

The most impressive study I still recall was about the adverse effects of antiepileptic drugs during pregnancy on mothers and later on their newborn infants. A retrospective investigation in subjects under treatment at the Department of Neuropsychiatry of Nagasaki University Hospital, which was conducted immediately after neonatal deformity expression in infants in maternal patients given an antiepileptic drug was reported at a Nagasaki Psychiatry Forum organized by local psychiatrists in 1971, revealed that issues associated with epilepsy treatment were more extensive than initially expected. Of 77 pregnancies while on antiepileptic drugs in 41 female epilepsy patients, 48 showed live births, 18 spontaneous abortions or stillbirths, and 11 artificial abortions. Of 48 productions, 6 (12.5 %) involved

Y. Nakane, *Progress in Social Psychiatry in Japan: An Approach to Psychiatric Epidemiology*, DOI 10.1007/978-4-431-54103-5_4, © Springer Japan 2012

congenital abnormality in newborn infants, 2 (4.2 %) hemorrhagic diseases, and 2 (4.2 %) were born prematurely [1]. Since around the mid-60s, a bit earlier than the implementation of the above investigation, results from some relevant studies have been reported. In particular, results from a multiple case study analysis presented by Dieter Janz, Professor at Berlin Free University (Freie Universität Berlin), Germany, have given a rise to a gradual increase in global interest on treatment with antiepileptic drugs during pregnancy [2].

When I came across an opportunity to present the investigation results from Nagasaki University at a current nationwide epilepsy forum, I was given a reproof by a professor of authority in epilepsy study for the following: *I've never had such experiences with the cases presented. Epilepsy patients under treatment should not be made to feel unsettled through treatments with results obtained from case studies with minimum samples. Such results can be obtained accidentally, so results based on multiple studies should be worth being discussed here.* When I saw him again after about three months, however, he modified his last comment stating: *After that day, I reviewed my cases and found some cases of neonatal deformity. We are urged to have a rigorous investigation.* Since then, this study was initiated and evolved into a large-scale, nationwide investigation in 1974, which was encouraged to continue with Grants for Science from the Ministry of Education in 1975 and fellowship grants from the Ministry of Health and Welfare from 1975 to 1977. For the national study with participation of 11 centers, 948 pregnancies in female epilepsy patients were recruited, and outcomes of compliance as well as noncompliance with antiepileptic drugs were summarized. In 1978, results were reported at the 11th Congress of the Japan Epilepsy Society held in Atami and at the 10th Epilepsy International Symposium held in Vancouver, Canada, and published in journals such as the *Japanese Journal of Brain and Nerves* [3] and *Epilepsia* [4]. Our achievement was recognized, and I was invited to the Epilepsy Seminar held in then West Germany in September 1980, hosted by Professor Janz.

One of the coresearchers, Sunao Kaneko, Professor at Hirosaki University, then moved on to the second stage of the investigation involving some foreign researchers, and their investigation results have been adopted as current guidelines for treatment with antiepileptic drugs during pregnancy.

In tandem with my personal engagement in epilepsy research, I have also worked on epidemiological research on adolescent mental disorders. Starting with epidemiological studies on a wide range of mental retardation, relevant activities were expanded to include research on Japan-specific clinical manifestation of Down syndrome (DS), surveys on psychological loads on families with a DS child, as well as supportive activities for the Kobato-kai, an association of parents with a DS child. For autism and Asperger's syndrome, reviews of epidemiological data were repeated, and, concurrently, new approaches to treatment were explored based on the symptomatologies and etiology. When childhood depression and hyperactive children aroused my interest, an investigation was conducted to identify environmental causalities at school. When my long-term experience with part-time service at a local children's counseling center in Nagasaki further extended my interest to concerns related to adolescent mental health in schools, including dealing with the problems of isolation from society and domestic violence, I officially took a position as a school doctor at a prefectural high school.

However, I can say that the backbone of our research emerged from research activities in the WHO Collaborating Center. Today, we have commemorated the 30th anniversary of becoming an official WHO Collaborating Center for Research and Training since its designation on August 31, 1979. I personally appreciated being able to participate in relevant research as a member of the WHO Advisory Expert Panel. About two years after the designation by WHO, we were allowed to celebrate the opening of our Collaborating Center. I would like to describe our opening ceremony in brief.

In order to bring a Collaborating Center designated by WHO into effective operation, it is recommended to set up supportive advisory committees for our activities. Such committees are categorized into the following three different levels: local advisory committee (organized by expert panels from local institutes such as Nagasaki University), national advisory committee (organized by domestic expert panels), and international advisory committee (organized by international expert panels and those who are in seamless contact with the head of the center).

The opening ceremony and the special lectures were held on February 26 and 27, 1981, with those named to the list of expert panels. On the day of the ceremony, national flags and the WHO flag were raised at a rotary in front of the Nagasaki University Hospital (Fig. 4.1).

In the late afternoon of February 24, Pope John Paul II arrived in Nagasaki for his first ever visit to Japan, with the intention of holding the Nagasaki Martyrs' Memorial Mass, which occasionally coincides with the time when our opening

Fig. 4.1 The rotary in front of the Nagasaki University Hospital on the day of the opening ceremony

THE JOINT MEETING OF ADVISORY COMMITTEES AT THREE LEVELS (INTERNATIONAL, NATIONAL AND LOCAL) AT THE WHO COLLABORATING CENTER FOR RESEARCH IN FUNCTIONAL PSYCHOSES, DEPARTMENT OF NEUROPSYCHIATRY NAGASAKI UNIVERSITY SCHOOI OF MEDICINE, FEBRUARY 26, 1981

Fig. 4.2 The joint meeting of advisory committee (1981). *From right of the front row*: Prof. Kusuo Sakurai (former Prof. at Kyusyu Univ.), Prof. H. Häfner (Mannheim), Prof. E. Strömgren (Aarhus), Dr. Hiroshi Nakajima [Director-General of the WHO Western Pacific Regional Office (WHO/WRPO)], Prof. Ryo Takahashi (Nagasaki Univ.), Prof. N. Sartorius (WHO, Chairman of the Mental Health Department), Prof. G. Klerman (Boston), Prof. Haruo Akimoto (former Prof. at Tokyo Univ.), Prof. Tadashi Inose (former Prof. at Yokohama Univ.) *From right of the back row*: Prof. Isao Fujii (Nagasaki Univ.), next person but one: Prof. Yasuo Shimazono (former Prof. at Tokyo Medical and Dental Univ.), Prof. Hiroshi Utena (former Prof. at Tokyo Univ.), Prof. Masaaki Kato (former Director of the National Institute of Mental Health), next person but two: Prof. Hideo Fukumi (President of Nagasaki Univ.), Prof. C. Leon, Dr. Nozomu Kawashima (division manager of the Health Department, Nagasaki Prefectural Government), Prof. Hirofumi Suyama (President of Nagasaki Univ. School of Medicine), Prof. Teruo Okuma (former Prof. at Tohoku Univ.), Prof. Takeo Doi (former Prof. at Tokyo Univ.), Prof. Shen Yucun (Beijing), Prof. Hideo Hosaki (former Prof. at Keio Univ.), Dr. Hideo Shinozaki (Ministry of Health and Welfare), Dr. Yoshibumi Nakane (Author), Dr. Yasuyuki Ohta (Nagasaki Univ.), and others (parentheses indicate their position at that time)

ceremony began. On February 26 in the morning, an unusually amount of snow fell in Nagasaki and completely disrupted transportation. Despite such conditions, 57,000 people gathered, and the mass was performed with stateliness.

On the other hand, our opening ceremony started a bit behind schedule, but, as shown in Fig. 4.2, it was completed safely and successfully in the presence of quite a lot of advisory committees.

The congratulatory speeches were given by guests of honor, including a representative from the Ministry of Health and Welfare, the governor of Nagasaki Prefecture, the mayor of Nagasaki City, and foreign researchers at the ceremony. The keynote lecture by Professor N. Sartorius (Fig. 4.3) and the commemorative speech by Hiroshi Nakajima, director-general of WHO/WPRO, were followed by special lectures given by Professor E. Strömgren, Professor G. Klerman, and some

Fig. 4.3 The latest photo of Professor Norma Sartorius (I took it in February 2010 in Tokyo)

Table 4.1 Special topics: "Social Psychiatry and Functional Psychoses" and "Mental Health Problems in Asia" [5]

Introduction	Ryo Takahashi, Nagasaki University
Commemorative speech	Hiroshi Nakajima, WHO Western Pacific Regional Office (WPRO)
Keynote lecture	N. Sartorius, the Department of Mental Health of WHO
The question of so-called "acute psychoses"	Erik Strömgren, Aarhus University (Denmark)
Neurotic and psychotic depression: a reevaluation	Gerald L. Klerman, Harvard University (US)
The impact of an extending community mental health service on patterns of bed usage (evaluation of a 4-year-period of implementation)	H. Häfner and J. Klug, the Central Institute of Mental Health (W. Germany)
Researches in mental health in South East Asia	Narendra N. Wig, All-India Institute of Medical Sciences (India)
New directions in cross-cultural mental health	Anthony J. Marsella, WHO/NIMH Schizophrenia Research
Psychiatric services in the urban and rural areas of the People's Republic of China	Shen Yu-cun and Zheng Wei-Xi, Beijing Medical University (China)
Mental health activities of WHO in the Western Pacific Region	Hideo Shinozaki, Department of National Hospitals, Health Service Bureau, Ministry of Health and Welfare

advisory committee members. As shown in Table 4.1, parts of these lectures have been translated into Japanese and published in a special edition of the *Clinical Psychiatry* [5].

Table 4.2 summarizes international comparative and collaborative studies our department has been working on with WHO/HQ for more than 30 years since 1972, before designation as the collaborating center.

The mental health division of WHO/HQ (initially the Division of Mental Health, currently the Division of Mental Health and Prevention of Substance Abuse)

Table 4.2 Summaries of WHO collaborative studies in the Nagasaki Collaborating Centers

1. Mood disorder studies
   - Studies on Standardized Assessment of Depressive Disorder; SADD; five centers, 1972–1979
   - 5-year, 10-year, and long-term follow-ups in patients selected for the standard evaluation; four centers, 1989–1994
   - Studies on Depressive Patients in General Practice; GP study; two centers, 1977–1980
   - Studies on Dose Effects of Antidepressant Medication in Different Populations; five centers, 1981–1985
   - Study on Comparison of Oral and Intravenous Treatment of Depressive States; 11 centers, 1981–1985
   - Study on the Usage of the Dexamethasone Suppression Test as a Biological marker for Depressive Disorder; 13 centers, 1982–1987
   - Comparative Studies of Affective Disorders among Three Asian Countries; three centers, 1985–1991
2. Schizophrenia studies
   - The Collaborative Study of Determinants of Outcomes of Severe Mental Disorders; DOSMeD
     - The core study: the study on detection of patients initially developing schizophrenia (the incidence rate study); 11 centers, 1976–1986
     - Relevant study 1: the study on susceptibility in patients with a complication of malignant tumor; five centers, 1979–1981
     - Relevant study 2: study on life events; nine centers, 1979–1997
     - Relevant study 3: study on emotional interaction of patients and the family and short-term outcomes (the Nagasaki Center did not participate)
     - Relevant study 4: study on social dysfunction (the Nagasaki Center did not participate)
     - Relevant study 5: study on acute psychosis (the Nagasaki Center did not participate)
   - International Study of Schizophrenia; ISoS; 18 centers, 1989–1997
3. ICD-10 (F) relevant studies
   - CDDG domestic collaborating studies; five centers, 1987–1989
   - DCR domestic collaborating studies; 28 centers, 1990–1991
   - PHC domestic studies; the Nagasaki Center
4. WHO Collaborative Study on Psychological Problems in General Health Care Settings; PPGHC
   - Study on treatment; 15 centers, 1990–1996
   - Study on sleep disorders in general practice; several centers, 1995–1998
5. Quality-of-life (QOL) studies
   - Study on development of WHO/QOL; multicenter, 1992–1997
   - QOL study in patients with cancer; two domestic centers and three overseas centers, 1995–1997
   - QOL study in caregivers for patients with dementia; four domestic centers, three overseas centers, 1996–2000
6. International Classification of Functioning, Disability, and Health; ICF
   - Study on development of the Japanese version; multicenter, 1996–2002
7. WHO/World Mental Health Survey in Japan; WMH-J; 17 countries, since 2002 (6 centers in Japan, 2002–2006)

commonly compares developed and developing countries (low- and middle-income countries vs. high-income countries) as one of the social psychiatric approaches to studying psychiatric disorders. It also has advanced, highly controlled studies based on examination and practice assessment approaches using predetermined scales, aiming to eliminate international differences in symptom assessment and diagnosis. It takes the perspective that proper social psychiatric examination of symptom assessment is possible in clinical practice if the basic background with many of its variances is well adjusted. Most assessment scales for such situations were developed in western countries, and few of those countries have a Japanese version. Therefore, many scales were introduced to Japan for the first time, and we had to translate them completely into Japanese before participating in the international collaboration, which was often very difficult for us. For example, the Japanese version of the *Standardized Assessment of Depressive Disorders* (*SADD*) [6] was developed as a scale for depression and has been used in subsequent clinical studies. Before initiating schizophrenia-related studies, we developed many survey forms for much more than depression-related studies. As described later, we planned

Table 4.3 Assessment scales used for the WHO collaborative studies

For mood disorder studies
1. Screening Schedule
2. Standardized Assessment of Depressive Disorder
3. Follow-up Screening Schedule
4. Screen Form for General Practitioners
5. Depression Screen for Psychiatrists
6. Hamilton Psychiatric Rating Scale for Depression
7. Rating Scale for Side Effects
For schizophrenia studies
1. Assessment scales used for the initial assessment
1) Screening Schedule
2) Psychiatric Personal History Schedule, PPHS
3) Diagnostic and Prognostic Schedule, DPS
4) Present State Examination-9th revision, PSE-9
5) Katz Adjustment Scale, KAS
6) Life Event Schedule, LES
2. Assessment scales used for the DOSMeD long-term follow-up
1) Present State Examination-9th revision, PSE-9
2) Psychological Impairment Rating Schedule II, PIRS-II/WHO
3) Scale for the Assessment of Negative Symptoms, SANS
4) Global Assessment of Functioning Scale for Symptoms, GAF-S/WHO
5) Disability Assessment Scale, DAS/WHO
6) Global Assessment of Functioning Scale for Disability, GAF-D/WHO
7) Life Chart Schedule, LCS/WHO
For studies on treatment of psychological problems in general practice
1. Screening Schedule
2. General Health Questionnaire-12 items, GHQ-12
3. Composite International Diagnostic Interview, WHO/CIDI, ver. 2
For Quality of Life (QOL) studies
1. Screening Schedule
2. General Health Questionnaire-12 items, GHQ-12
3. WHO/Quality of Life-Brief, WHO/QOL-26
International collaborative studies on mental health
1. Screening Schedule
2. Composite International Diagnostic Interview, WHO/CIDI, ver.3, CAPI; parts I and II

schizophrenia-related studies to have some core studies and many ancillary studies, and individual researchers needed to participate in the development of assessment scales for the study they would be involved in. Many scales were developed through subsequent collaborative studies, and some of them are now being commonly used in subsequent research studies.

Table 4.3 shows the list of assessment scales adopted in studies of depression, schizophrenia, general practice, quality of life (QOL), and mental health surveys (in the WHO World Mental Health survey initiative). We developed the Japanese versions based on the English version of the WHO International Statistical Classification of Diseases and Related Health Problems and the WHO International Classification of Functioning, Disability, and Health.

A researchers' meeting was held before initiating studies of schizophrenia. The commemorative photo taken at the meeting (Fig. 4.4) always reminds me of the heated debates that went on for several days among researchers in specialized fields.

PSE-9 (Present State Examination) of the aforementioned scales, which I describe in detail later, was revised and included in PSE-10, which allows symptom assessment of more psychiatric disorders as part of the Schedule for Clinical Assessment in Neuropsychiatry (SCAN). Also, the WHO's Disability Assessment Schedule was improved to become the International Classification of Functioning, Disability, and Health (ICF), showing promising use after ICIDH (the International Classification of Impairments, Disabilities, and Handicaps), ICIDH-2, and ICF. The Composite International Diagnostic Interview (CIDI) was improved to the computer version, the Computer-assisted Personal interview (CAPI) after repeated revisions, and it is

Fig. 4.4 The first DOSMeD researchers' meeting (WHO/HQ in Geneva, Switzerland, 1977.6.22) Dr. M. Fischer, the second woman from the right in front; Dr. N. Sartorius in the black suit behind her; Dr. E. Strömgren, left rear from Dr. Sartorius; the author, behind the fourth woman from the left in front; Dr. J. Cooper, left rear from the author; Dr. J. Leff, behind Dr. Cooper; Dr. L. Wynne, a tall man on the right; Dr. M. Katz, in front of Dr. Wynne

compatible with the diagnosis classification system of either ICD or DSM. Looking back on these processes, WHO-led international collaborative research will play a role as an international lingua franca in future.

References

1. Nakane Y, Takahashi R. Associations between anti-epileptic drugs administered during pregnancy and abnormalities of the children. Jpn J Clin Psychiatry. 1974;3:363–71 (in Japanese).
2. Janz D, Fuchs U. Sind antiepileptische Medikamente während der Schwangerschaft schädlich? Deutsche Medizinische Wochenschrift. 1964;89:241–8.
3. Okuma T, Takahashi R, Nakane Y, et al. Teratogenic effect of antiepileptic drugs: a collaborative study in Japan. Adv Neurol Sci. 1979;23:1247–63 (in Japanese).
4. Nakane Y, Okuma T, Takahashi R, et al. Multi-institutional study on the teratogenicity and fetal toxicity of antiepileptic drugs: a report of a collaborative study in Japan. Epilepsia. 1980;21:663–80.
5. Sartorius N, Strömgren E, Klerman GL, et al. Memorial lectures for the opening of the WHO Collaborating Center for Research on Functional Psychoses. Clin Psychiatry. 1981;23:1092–169 (translated into Japanese).
6. Sartorius N, Davidian H, Ernberg G, et al. Depressive disorders in different cultures; report on the WHO Collaborative Study on Standardized Assessment of Depressive Disorders. Geneva: WHO; 1983.

Chapter 5
Studies on Schizophrenia

5.1 Incidence Rates, Prevalence Rates, and Other Statistics

As Table 5.1 shows, Japan's first study on the frequency of psychiatric disorders was conducted in 1940 in Hachijo-jima Island (Tokyo) by a group from Tokyo University led by Professor Yushi Uchimura [1]. This was a genetic epidemiological study with a sample of not only subjects with schizophrenia but also those with disorders categorized into major psychoses, including bipolar affective disorders, following the preceding study initiated by C. Brugger in Thüringen, Germany, in 1931. Every possible measure is considered to have been taken to obtain as much appropriate information as possible from accessible networks in particular areas. This study was recognized as a pioneering approach and followed by a number of regional studies, as summarized in Table 5.2. The focus of these studies was to determine lifetime prevalence, and the study results were used to estimate morbidity risk following Weinberg's convention. Since this was a commonly used method at that time and the deviations among each of the morbidity risk rates shown in the bottom of Table 5.2 are within an acceptable range, estimates are considered reliable for comparison of data to an extent, though coverage accuracy and inclusion criteria of the sample are not clearly given.

Table 5.3 compares data from Asian countries excluding Japan to those from other foreign countries, as well as data extracted from our study results on schizophrenia incidence rates which we have published on a regular basis in literature such as *A Prospective Study on the Relationship between Physical* [2] *and Mental Development in Childhood and Psychiatric Disorders in Later Life* [3] and *A Study of Schizophrenia among Atomic Bomb Survivors* [4], in addition to the DOSMeD (Determinants of the Outcome of Severe Mental Disorders) study, into which we will delve into in detail later on. Even though small differences were observable among the data shown in this table, their level has been deemed tolerable.

During the review process of differences identified in frequency, the consistency in criteria used for grouping the sample in all studies was questioned, that is, the necessity of controlling the concepts of disorders was brought into the discussion.

Y. Nakane, *Progress in Social Psychiatry in Japan: An Approach to Psychiatric Epidemiology*, DOI 10.1007/978-4-431-54103-5_5, © Springer Japan 2012

Table 5.1 History of prevalence studies in Japan [1]

Researcher	Research areas (year)	Population at that time	Narrow psychiatric disorder
Uchimura et al.	Hachijojima (1940)	8,313	0.68 %
Akimoto et al.	Hachijojima (1960)	12,027	0.88
Uchimura et al.	Miyakejima (1940)	5,286	0.91
Akimoto et al.	Komoro (1941)	5,207	1.08
Tsugawa et al.	Ikebukuro (1940)	2,713	1.22
	All areas	21,518	0.90
Mukasa et al.	Gokanosyo (1940)	1,322	0.61
Hirayasu	Henzajima (1965)	2,379	5.04
Arai et al.	Saitama T (1957)	2,464	0.57
Ministry of Health and Welfare	Nationwide (1963)	44,092	0.59

Accordingly, it was decided to conduct a comparative study of diagnostic processes of psychiatric disorders similar to the earlier mentioned projects in the USA and the UK involving researchers in western countries consisted mainly of WHO experts, to establish diagnostic criteria for particular disorders based on the study results. Among those who participated in this study, the American Psychiatric Association (APA) has played a pilot role within the relevant scope in an attempt to develop the related diagnostic interview schedule (DIS) through the establishment of the DSM (Diagnostic and Statistical Manual) system.

After that, the DSM system produced by the APA was widely used in epidemiological and regional studies and eventually became a global trend. In conjunction with this, application of the DISs, including PSE (Present State Examination), SCAN (Schedules for Clinical Assessment in Neuropsychiatry), and CIDI (Composite International Diagnostic Interview), based on the ICD (International Classification of Diseases) system created by WHO, became common in regional studies. Turning a deaf ear to this international trend, Japan fell behind other Asian countries in the implementation of similar studies due to interview schedules and assessment instruments being available only in English. In addition to these circumstances, there might still have been a lack of proper case-finding methodologies, etc., such as assessment scales, to have a clear vision for the future. At any rate, the global trend was about to shed the conventional genetic epidemiological instruments in support of the social epidemiological instrument.

According to the accumulated findings, and eliminating the overlap with other review papers we have already published, the prevalence rate and incidence rate of schizophrenia can be summarized as follows [5–7].

For quite a while, schizophrenia had been considered to occur at a similar rate globally (although exceptionally high rates of cases were also known). An analytical review of about 200 articles extracted from more than 1,700 articles published between 1965 and 2002 [8] concluded, however, that results had shown not significant but small inter-study gaps in the mean point prevalence rate of 0.64 % (minimum 0.19, maximum 1.00, a fivefold difference) and mean lifetime prevalence rate of 0.40 % (minimum 0.18, maximum 1.16, a 6.4-fold difference).

Table 5.2 The morbidity risk in patients with schizophrenia observed in the prevalence study in Japan (%)

Researcher (year)	Research area	Population	Subjects	Patients	Morbidity risk[a]	Morbidity risk[b]
Uchimura et al. (1940)	Hachijojima in Tokyo	Total 8,318	3,081	32	0.91	1.04
		Male 3,998	1,411	22		1.56
		Female 4,320	1,670	10		0.60
Mukasa et al. (1941)	Gokanosyo in Kumamoto, etc.	Total 4,443			0.74	
Hiratsuka et al. (1941)	Muraoka in Kanagawa	Total 1,704	634	7	0.98	1.10
		Male 851	309	4		1.29
		Female 853	325	3		0.92
Tsugawa et al. (1942)	Ikegukuro in Tokyo	Total 2,712	1,080	6	0.49	0.56
		Male 1,371	553	3		0.54
		Female 1,341	527	3		0.57
Uchimura et al.(1942)	Miyakejima in Tokyo	Total 5,284			0.64	
Akimoto et al. (1943)	Komoro in Nagano	Total 5,207	1,999	11	0.50	0.55
		Male 2,470	898	6		0.67
		Female 2,737	1,023	5		0.49
Ogino et al. (1943)	Bouzejima in Hyogo	Total 1,651			0.52	
Nomura et al. (1944)	Hinohara in Tokyo	Total 1,758			0.82	
Okabe (1957)	Koshikijima in Kagoshima (consanguineous area)	Total 6,783	2,657	50	1.68	1.88
		Male 3,206	1,209	29		2.40
		Female 3,557	1,448	21		1.45
Arai et al. (1958)	Chichibu Kuranama in Saitama	Total 2,401			2.47	
Arai et al. (1958)	Chichibu Tomita in Saitama	Total 2,464			0.48	
Arai et al. (1959)	Ukishima in Ibaraki	Total 2,403			0.91	
Akimoto et al. (1964)	Hachijojima in Tokyo	Total 12,027	5,054	57	1.03	1.13
		Male 5,941	2,331	32		1.37
		Female 6,086	2,473	35		1.01

(continued)

Table 5.2 (continued)

Researcher (year)	Research area	Population	Subjects	Patients	Morbidity risk[a]	Morbidity risk[b]
Hirayasu (1969)	Henzajima in Okinawa	Total 2,379	916	21	2.11	2.29
		Male 1,104	398.5	10		2.51
		Female 1,275	517.5	11		2.13
Haruki (1972)	Tsuma, Oki-district in Shimane	Total 2,826	1,444	24	1.07	1.66
		Male 1,385	672	12		1.78
		Female 1,441	772	12		1.55

The number of subjects = $b-(\mathrm{bo}+\mathrm{bm})$; b: population in the research area; bo: the number of people who did not reach the age at risk of schizophrenia; bm: the number of people of which ages were at risk of schizophrenia

[a]Morbidity risk (age 16–40) provided by the original author

[b]Morbidity risk (age 15–54) recounted by Nakane et al.

Table 5.3 Typical studies on prevalence rate and incidence rate of schizophrenia (overseas data) [4]

Reporter	Catchment area	Prevalence rate and incidence rate	Age	Morbidity risk (%)
Reports from Asian countries other than Japan (prevalence rate)				
Lin (1953)	Taiwan	2.2 (lifetime)	All ages	
Yoo (1961)	Korea	3.8 (lifetime)	All ages	0.96
Jayasundera (1969)	Ceylon	3.2 (at certain point)	All ages	
Lin (1969)	Taiwan	1.4 (lifetime)	All ages	
Dube (1970)	India	3.7 (lifetime)	All ages	0.60
Wijesinghe (1978)	Ceylon	3.8 (certain period)	15 years and up	0.88
Shen (1981)	Beijing	1.8 (certain period)	All ages	0.46
Rin (1982)	Taiwan	0.9 (lifetime)	All ages	
Typical annual incidence rate studies				
Odegaad (1946)	Norway	2.4	10 years and up	1.87
Pollak (1954)	Maryland	6.3	15 years and up	
	Louisiana	5.7	15 years and up	
Hollingshead (1958)	New Haven	3.0	15 years and up	
Norris (1959)	London	1.7		
Jaco (1960)	Texas	3.5	15 years and up	
Dunham (1965)	Detroit/high	12.0	15 years and up	
	Detroit/low	4.5	15 years and up	
Warthen (1967)	Maryland	5.0	All ages	
Adelstein (1968)	Salford	3.3	15 years and up	
Walsh (1969)	Dublin	5.2	10 years and up	
Babigian (1970)	Monroe county	6.9	15 years and up	M 0.47 F 0.31
Hafner (1970)	Mannheim	5.4	All ages	
Lieberman (1974)	Moscow	1.9	All ages	

(continued)

Table 5.3 (continued)

Reporter	Catchment area	Prevalence rate and incidence rate	Age	Morbidity risk (%)
Babigian (1975)	Monroe county	5.5	15 years and up	M 0.35 F 0.24
Helgason (1977)	Iceland	2.7	15 years and up	M 0.43 F 0.54
Gibbons (1981)	Southermpton	2.5	15 years and up	
Robertson (1982)	Aberdeen	1.9	15 years and up	
Wing (1984)	Camberwell	1.4	15 years and up	M 1.19 F 1.25
Goldstein (1984)	New South Wales	3.0	15 years and up	M 0.63 F 0.59

However, unlike the assumption, almost no gap in frequency of point prevalence and lifetime prevalence has usually been pointed out between studies. In terms of risk factors associated with the onset of schizophrenia, topics such as higher prevalence rates in men than women (about 1.4-fold higher), urbanization and immigration (immigrants), single households, high latitudes, birth during winter, and perinatal complications are increasingly drawing closer attention.

5.2 Findings of the WHO Collaborating Center in Nagasaki

The WHO Collaborating Center in Nagasaki joined the WHO Collaborative Study on Determinants of Outcome of Severe Mental Disorder, which has been ongoing since 1971 and has been monitoring subjects in certain regions over a long period.

DOSMeD is a study that evolved from the International Pilot Study of Schizophrenia (IPSS) initiated in 1966 (see Table 5.4) [9]. The main aim of the DOSMeD study is to verify the appropriateness of findings obtained from the IPSS. Study results revealed that schizophrenia patients were identified with a similar frequency in all the nine participating countries, but, contrary to the estimates, a clinically and socially better outcome was obtained with the subjects in developing countries than those in developed countries at 2 and 5 years after inclusion in this study. We were required here to address another issue, as various opinions about the appropriateness of establishing the living area, which was categorized into developing countries and developed countries, as a predictor of outcome, including possible bias and errors in results.

Today, nearly 30 years after the DOSMeD study was initiated, almost all of the co-researchers who started this study together have already left Nagasaki University Hospital, which served then as the Collaborative Center. They are still involved in medical care, some at psychiatric hospitals and others at their own medical offices.

Table 5.4 Centers participating in IPSS and DOSMeD [9]

	Centers in developed countries	Centers in developing countries
IPSS	Aarhus (Denmark) Moscow (USSR) London (United Kingdom) Prague (Czechoslovakia) Taipei (China) Washington (USA)	Agra (India) Cali (Colombia) Ibadan (Nigeria)
DOSMeD	Aarhus (Denmark) Dublin (Ireland) Honolulu and Rochester (USA) Moscow (USSR) Nagasaki (Japan) Nottingham (United Kingdom) Prague (Czechoslovakia)	Agra and Chandigarh/R/U (India) Cali (Colombia) Ibadan (Nigeria)

Table 5.5 The inclusion criteria used in the DOSMeD study (criteria for psychiatric symptoms and behavioral abnormalities)

1. During the past one year has the patient presented any of the following?
   (a) Hallucination or pseudohallucination in any modality
   (b) Delusions
   (c) Marked thought and speech disorder (e.g., incoherence, irrelevance, thought blocking, neologism, incomprehensibility of speech) other than simple retardation or acceleration
   (d) Marked psychomotor disorder (e.g., negativism, mutism or stupor; catatonic excitement; constrained attitudes or unnatural postures maintained for long periods) other than simple retardation or acceleration
   (e) Emergence or marked exacerbation of bizarre and grossly inappropriate behavior (e.g., talking or giggling to self, acts incomprehensible to others, and loss of social constraints)

Or

2. During the past one year, has the patient presented a definite *change* of personality and behavior manifested in any of the following?
   (a) Marked reduction or loss of interests, initiative and drive, leading to serious deterioration of the performance of usual activities and tasks
   (b) Emergence or marked exacerbation of social withdrawal (active avoidance of communication with other people)
   (c) Severe excitement, purposeless destructiveness or aggression
   (d) Episodic or persistent states of overwhelming fear or severe anxiety
   (e) Gross and persistent self-neglect

As time goes by, not only the study subjects, but also some study staff members involved have changed their status. To my disappointment, some of the senior researchers (e.g., Professor Ryo Takahashi) and peers have passed away.

Now I would like to give some detailed explanation on the subjects that were included and the methods that were employed in the DOSMeD study.

Study subjects were selected according to inclusion/exclusion criteria specified in the predefined screening schedule (SS). The patient sample included in this study consisted of a Nagasaki-based cohort in a catchment area for at least 6 months retrospectively from the study start, aged 15–54, and who experienced the first onset of behavioral abnormalities and psychotic symptoms during the study period. Those who were intellectually disabled to the extent that it disturbed the collection of appropriate information (having an IQ score of less than 36 or clinically the same level) were excluded. As clear exclusion criteria, subjects should not be afflicted with clinically significant organic brain disorder, such as infectious, parasitic, addictive, cerebrovascular, and epileptic encephalopathy, or significant encephalopathy involving organic central nervous system disorder and polyneuropathy caused by serious abuse of alcohol or drugs. For inclusion criteria or *behavioral abnormalities and psychotic symptoms*, refer to Table 5.5.

Before entering into the main stage of this study, a 2-month pilot study was conducted in October and November 1978 to identify the contact institutions of the subjects meeting the above inclusion criteria. Study results revealed that some subjects sought consultation at psychiatric hospitals in Nagasaki City as well as in surrounding areas, and at not only psychiatric hospitals but also service institutions such as a public health center. Following this, we asked 30 institutions in total—18 private psychiatric hospitals, 1 prefectural psychiatric hospital, 5 private psychiatric practices, 4 psychiatric departments in national general hospitals, and 2 public health centers—for their cooperation as case-finding networks for at least 2 years.

The main activity of this case-finding phase took place from January 1, 1979, through December 31, 1980. During this time period, one research-assistant of the Collaborating Center was usually involved alone in a telephone survey of all 30 institutions daily at 11 a.m. (on weekends or national holidays, the day after that). The content of the survey was designed by the Collaborating Center team as follows. After greeting, request for presence of first consultation with mental disorder outpatients on the day of phone call or within days, and, if any, the inclusion possibility of the patient in this study was requested (see Table 5.5). If the case was considered to satisfy the inclusion criteria, the possibility of having an interview with the physician for outpatient treatment at the relevant institution was requested. When the possibility was confirmed, the attending clinician or physician for outpatient treatment was then requested to provide the patient and his or her family with a brief explanation about the main purpose and content of this study, and when agreement was tentatively reached, a researcher from this study visited the relevant institution sometime after that. When the patient and his or her family were able to understand the consequences of participation and consented to be included in this study, fulfillment was determined at that time. Once accepted, the interview with the patient and his or her family was started for thorough assessment.

The case-finding process of this study is illustrated in Fig. 5.1. As seen in the figure, another 2-year leakage study (with an intent to check leakage of relevant

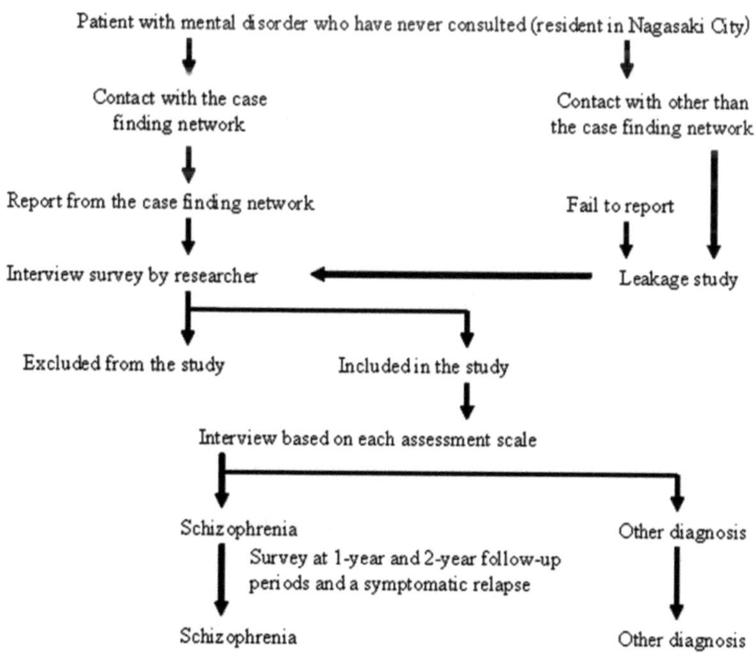

Fig. 5.1 DOSMeD case finding system

Table 5.6 Monthly distribution of the patients reported during the 2-year research

	Reported cases			Accepted cases			Schizophrenia		
Month	Male	Female	All	Male	Female	All	Male	Female	All
January	9	6	15 (2)	7	4	11	6	4	10
February	7	7	14 (1)	5	6	11	5	4	9
March	15	4	19 (1)	10	1	11	9	1	10
April	13	10	23 (1)	8	5	13	6	4	10
May	4	5	9 (1)	2	5	7	2	4	6
June	7	5	12	4	3	7	4	3	7
July	7	5	12 (1)	7	3	10	6	2	8
August	7	3	10	4	2	6	4	2	6
September	4	11	15	2	9	11	2	7	9
October	9	12	21 (3)	6	10	16	5	9	14
November	5	6	11	4	4	8	4	4	8
December	9	3	12	8	2	10	8	2	10
Total	96	77	173 (10)	67	54	121	61	46	107

information from the case-finding networks' reports and to identify institutions other than our corresponding networks that had psychiatric outpatients by referral data from the health center, etc.) followed the initial 2-year study. In sum, this regional DOSMeD study spanned more than 6 years from the first co-researchers' meeting in June 1977 through to the completion of the leakage study at the end of 1982. The reason for setting a 2-year period for the leakage study was based on the result of an investigation which revealed that the maximum time in Nagasaki from symptom expression to the first consultation at psychiatric hospitals was 24 months.

Table 5.6 summarizes this information, including report examples, included subjects, as well as monthly and gender distributions of the cases diagnosed with schizophrenia received from the designated case-finding networks and other institutions over two years, including the leakage study period (values in parentheses are findings in the leakage study). One hundred and seventy-three cases were reported from the case-finding networks during the 2 years; 121 of these (69.9 % of all cases reported) met the inclusion criteria for this study, and 107 of these (61.8 % of all cases reported, 88.4 % of the fulfilled) were clinically diagnosed with schizophrenia. This table suggests neither specific patterns in the monthly distribution, seasonal increase of patients, for example in spring, nor change in number of patients by gender in certain months (reported slightly more cases in women in September and October, though) throughout the two years. On average, 5 (varying between 3 and 7) outpatients were diagnosed with schizophrenia monthly, totaling more than 50 annually.

Table 5.7 shows the distribution of schizophrenia by age group at the midpoint of the given study period (December 31, 1979) in the general population of the catchment area (population aged between 15 and 54 in Nagasaki) as a statistical background. Figures in the right-most column indicate incidence rates per 10,000 people according to gender and age. The total population in Nagasaki City at the given

Table 5.7 Population at the median period of the study and the annual incidence rate for schizophrenia by sex and age

	General population[a]			Schizophrenia			Incidence rate[b]		
Age	Male	Female	All	Male	Female	All	Male	Female	All
15–19	17,539	17,380	34,919	22	9	31	6.3	2.6	4.4
20–24	15,558	19,320	34,879	12	17	29	3.9	4.4	4.2
25–29	18,856	21,156	40,012	15	8	23	4.0	1.9	2.9
30–34	17,999	19,241	37,240	6	4	10	1.7	1.0	1.3
35–39	14,458	16,354	30,812	3	3	6	1.0	0.9	1.0
40–44	13,769	15,812	29,581	3	3	6	1.1	0.9	1.0
45–49	14,185	16,042	30,227	0	1	1	–	0.3	0.2
50–54	12,945	14,828	27,773	0	1	1	–	0.3	0.2
All	125,305	140,133	265,442	61	46	107	2.4	1.6	2.0

[a]General population: as of December 31, 1979
[b]Incidence rate: per 10,000 people

time-point was 447,091, and, among these, the 15- to 54-year-old population, which was the target age group, was 265,442. Distribution by age group showed a difference in incidence rates between men and women. While incidence rates in the male population reached a peak of 6.3 in the late-teenage group between 15 and 19, rates in women reached a peak of 4.4 in the early twenties, five years later than the peak in men, and the incidence persisted into some later decades. A slightly higher incidence rate was observed in men (2.4) than women (1.6), and the annual incidence rate was 2.0 per 10,000 people.

Based on the incidence rate, morbidity risk (also referred to as disease expectancy or individual risk), which is frequently referred to, can be estimated. Since incidence data are extremely rarely available, morbidity risk has alternatively been calculated based on prevalence rate (lifetime prevalence rate is commonly used), following Weinberg's convention (see Table 5.2). Since details on Weinberg's convention are given separately, the method of calculating morbidity risk based on the incidence rates will here be shown according to DOSMeD data. If lifetime incidence rates are available for all age groups (e.g., at an interval of 5 years), morbidity risk (P) is calculated by summing all incidence rates. This will be expressed in the following formula:

$$P\,(\%) = 5 \times (i_1 + i_2 + i_3 + i_4 + i_5 + i_6 + \cdots + i_n)/100$$

i_1: incidence rate in minimum age group; i_2: incidence rate in the next older age group; i_3: incidence rate in the next older age group; i_4: incidence rate in the next older age group; i_5: incidence rate in the next older age group; i_6: incidence rate in the next older age group; i_n: incidence rate in maximum age group.

Taking an example from the Nagasaki data (see Table 5.7):

$$\begin{aligned} P\,(\text{Nagasaki}) &= 5 \times (4.4 + 4.2 + 2.9 + 1.3 + 1.0 + 1.0 + 0.2 + 0.2)/100 \\ &= 5 \times 15.2/100 = 0.76\,(\%) \end{aligned}$$

The derived incidence rate in both sexes combined in the 15- to 54-year-old population in Nagasaki is 0.76 %.

If we now compare the accumulated morbidity risk data based on the past prevalence study results to foreign reports, which reflect a compilation of data from Scandinavian epidemiologists, Strömgren et al. reported 0.63 % and 0.72 % (studies in 1938 and 1989 [10]), Helgason et al. 0.7 and 1.1 (studies in 1964–1989 [11]) and Hagnell et al. 2.1 and 0.7 (studies in 1966 and 1989 [12]) in men and women, respectively. On the other hand, men showed risks of 0.54–2.51 (median 1.37 %), women 0.49–2.13 (median 0.92 %), and both sexes combined 0.55–2.29 (median 1.10 %) in Japanese prevalence studies (1940–1972, see Table 5.2). Meanwhile, overall morbidity risk derived based on our incidence rate data was 0.76 % in both sexes combined, 0.90 % in males and 0.62 % in female. Following these data, comparison of accumulated domestic data suggests that morbidity risk based on incidence rate is somewhat smaller than estimates based on prevalence rate. This would imply, however, that the usual morbidity risk of 1 per 100 people may rather be an overestimate of the actual rate, and one should thus deem the rate of 7–8 per 1,000 people appropriate.

Subsequently, we will compare the Nagasaki data to those collected from the other 12 centers in the 10 DOSMeD study participating foreign countries. I have to mention in advance that the compiled data were not contributed to by all centers, as progress status varied from center to center, and the data reflect the results at a certain time point.

As Table 5.8 shows, while there is a significant difference among centers in schizophrenia incidence rates, from 0.16 to 0.42 per 1,000 people (a 2.6-fold difference between minimum and maximum values) when calculated based on the included subjects clinically diagnosed with schizophrenia, the difference in narrow schizophrenia incidence rates of 0.07–0.14 per 1,000 people among centers was

Table 5.8 The incidence rate (per 1,000 people) and morbidity risk for schizophrenia (%) in the WHO/DOSMeD study [10–12]

Catchment areas	Broad schizophrenia		Narrow schizophrenia	
Aarhus (Denmark)	0.16 (0.18/0.13)	0.59 (0.68/0.51)	0.07 (0.09/0.05)	0.27 (0.33/0.20)
Chandigarh/rural (India)	0.42 (0.37/0.48)	1.72 (1.48/2.03)	0.11 (0.13/0.09)	0.48 (0.54/0.40)
Chandigarh/urban (India)	0.35 (0.34/0.35)	1.10 (1.04/1.21)	0.09 (0.08/0.11)	0.30 (0.22/0.42)
Dublin (Ireland)	0.22 (0.23/0.21)	0.83 (0.85/0.80)	0.09 (0.10/0.08)	0.32 (0.31/0.32)
Honolulu (USA)	0.16 (0.18/0.14)	0.50 (0.55/0.47)	0.09 (0.10/0.08)	0.26 (0.27/0.26)
Moscow (Russia)	0.28 (0.25/0.31)	1.13 (1.08/1.17)	0.12 (0.10/0.14)	0.47 (0.39/0.54)
Nagasaki (Japan)	0.20 (0.24/0.16)	0.76 (0.90/0.62)	0.10 (0.11/0.09)	0.37 (0.39/0.34)
Nottingham (UK)	0.22 (0.28/0.15)	0.80 (0.98/0.62)	0.14 (0.17/0.12)	0.54 (0.60/0.47)

The incidence rate (in the *left column*) and the morbidity risk (in the *right column*) of broad/narrow schizophrenia are the total of males and females combined. While the frequencies of broad schizophrenia were determined by clinical diagnosis, the frequencies of narrow schizophrenia were based on the cases classified in the Catego class S+ (described later)

slightly reduced to twofold when derived according to Catego class S+ (details follow later in this chapter) using the PSE of the structured interview schedule. Looking at the difference in morbidity risks among centers, about a 3.4-fold difference is observable in broad schizophrenia rates, ranging from 0.50 to 1.72, and about a twofold difference in narrow schizophrenia rates ranging from 0.26 % to 0.54 %. In the case of broad schizophrenia, such a result might have conventionally been considered to be associated with the cross-cultural difference in concepts or schemes of diseases and clinical diagnosis. However, if a twofold difference is also, in fact, to be seen in the narrow schizophrenia cases, this may be considered to be showing a clear regional gap.

Farris and Dunham gave their insight into residential patterns of inpatients before first hospitalization in their book on schizophrenia *Mental Disorders in Urban Areas, an ecological study of schizophrenia and other psychoses* [14]. As Fig. 5.2 indicates, the most frequent hospitalization was related to inhabitants from necrotic hubs suffering extreme economic hardships and the decay of society. The difference in number of inpatients was significant when compared to (bipolar) mood disorder

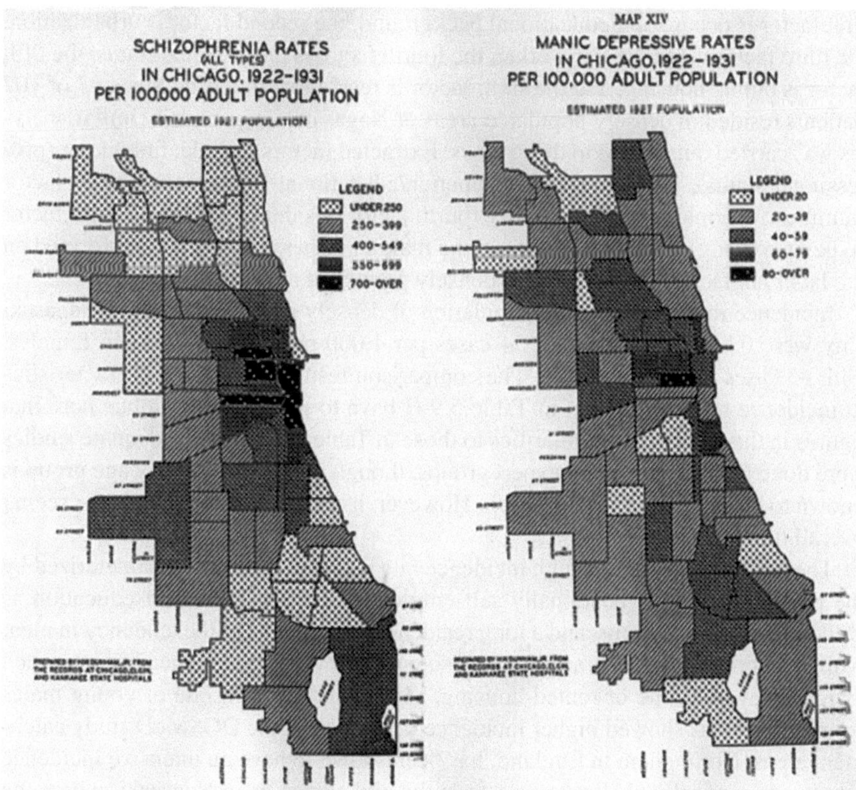

Fig. 5.2 Chicago Study (1939) Schizophrenia (*left*) and manic depressive rates (*right*) in Chicago 1922–1931 [13]

patients. Many similarly skewed distributions to theirs focused on schizophrenia have since been reported. Such a distribution pattern was interpreted into two hypotheses: *breeder hypothesis* (the low-income class is highly exposed to social and economical stresses that play a causative role in developing schizophrenia) and *social-selection-drift hypothesis* (the personality disorder was already seen before clinical expression and, as a result, was given a position with less responsibility and income). These are under debate as to causative meaning and afterthought. It was not an afterthought in a strict sense, however. The majority view appears to believe the latter hypotheses are proper.

Accordingly, we reviewed the 107 schizophrenia cases in Nagasaki by referring to opinions pertinent to regional characteristics from experts in public health, sociology, and related fields [15].

As a first step, data on the urban structure of Nagasaki City were re-aggregated based on regional data from the national poll in 1980 (44 input variables in total, including population, households, age structure, move-in date, previous residency, labor, vocational structure, educational background, residential setting, and others), and, through the application of factor analysis using Varimax orthogonal rotation, the following factors were predominantly extracted in the whole of Nagasaki. The first factor is occupation/educational background, the second factor is urbanization, the third factor is blue-collar worker, the fourth factor is professional status, the fifth factor is public housing, and the sixth factor is residential mobility. Since 97 of 107 patients resided in densely populated areas of Nagasaki City, another similar analysis was carried out limited to those areas. Extracted factors include: first factor (professional status), second factor (occupation/educational background), third factor (number of people in the household), fourth factor (residential mobility), fifth factor (type of housing), and sixth factor (young male). Further analysis was performed on the latter six factors extracted from densely populated areas exclusively.

Incidence rates in the whole population of densely populated areas in Nagasaki City were 0.500 in males, with 54 cases per 1,000 people, and 0.352 in females, with 43 cases per 1,000 people. The comparison results of regional characteristics to incidence rates are shown in Table 5.9 (I have to give advance notice here that figures in this table show similarities to those in Table 7.8, since no adequate studies were done in advance among expert groups, though comparing data by age group is known to be an appropriate approach. However, it seems to be acceptable for seeing overall trends).

The table suggests that a high incidence rate was shown in areas characterized by the population being vocationally self-employed, with overall lower education as well as professional status, and a long residency of habitants as the tendency in men, while in women, areas characterized by suburban, lower home-ownership rate, such as residency in public or rented housing, and higher move-in rate of young males from urban areas showed higher incidence rates. One of the DOSMeD study catchment areas, Nottingham in England, has been shown to have an intensive incidence rate in economically challenged strata in the city center, as in Nagasaki, where the same tendency was observed in socially disadvantaged areas, though not so significant as the cases in Nottingham. Following this, however, sociologists gave us

Table 5.9 Characteristics of the densely inhabited area in Nagasaki City and the incidence rate of schizophrenia

		Area characteristics	Incidence rate, males	Incidence rate, females
First factor	Professional status	Self-employed worker		
		Up to –1.0	0.870	0.135
		–1.0 to 1.0	0.491	0.397
		1.0 and up	0.391	0.142
		Employee		
Second factor	Occupational/educational background	Blue-collar		
		Up to –1.0	0.593	0.210
		–1.0 to 1.0	0.480	0.358
		1.0 and up	0.494	0.482
		White-collar		
Third factor	Household	Urban type		
		Up to –1.0	0.412	0.301
		–1.0 to 1.0	0.561	0.307
		1.0 and up	0.310	0.571
		Suburban type		
Fourth factor	Residential duration	At least 5 years		
		Up to –1.0	0.995	0.205
		–1.0 to 1.0	0.497	0.380
		1.0 and up	0.347	0.166
		Less than 5 years		
Fifth factor	Type of housing	Own house		
		Up to –1.0	0.394	0.346
		–1.0 to 1.0	0.499	0.339
		1.0 and up	0.524	0.438
		Public housing		
Sixth factor	Sex and age	Elderly female		
		Up to –1.0	0.202	0.165
		–1.0 to 1.0	0.522	0.360
		1.0 and up	0.571	0.450
		Young male		

Note: The incidence rates per 1,000 people

advice that a continuous study on the association of such residential characteristics with the development of disorders is necessary, since the effect of urbanization on the rate of mental disorders has also been reported.

5.3 Outline of Scales Used in the First Assessment or WHO/DOSMeD

In this DOSMeD study, the following scales were prepared for the assessment of clinically diagnosed schizophrenia cases, based on the screening schedule (SS) that specifies inclusion/exclusion criteria. As the initial assessment, a semi-structured interview study was conducted using: (1) the Psychiatric Personal History Schedule (PPHS), (2) the Diagnostic and Prognostic Schedule (DPS), (3) the Present State Examination-9th revision (PSE-9), (4) the Katz Adjustment Scale (KAS), and (5) the Life Event Schedule (LES).

In order to collect more accurate information from the clinical psychiatric viewpoint, appropriate assessment scales were newly developed, and scales with a demonstrated reliability in preceding studies were used.

PPHS aims at a systematic interview of the patient's past-history of disorders, DPS at specification of the diagnostic process and evaluation of the interviewer-oriented outcome prediction process, and LES at systematic identification of life events deemed relevant to onset and relapse of diseases. All the scales were developed further for this study. On the other hand, PSE-9 and KAS were already known as existing scales. The former aims to understand psychiatric symptoms precisely, and the latter to see the patients' adjustability to social life.

I would herewith like to give a brief explanation of the PSE. Development of this scale was begun by Wing et al. around 1962 by generating a comprehensive evaluation sheet with interview questions on psychiatric symptoms. It was proclaimed as PSE in 1967. After several revisions, PSE-9 was introduced under the names of the collaborative researchers Cooper, Sartorius, as well as Wing as the Semi-structured Interview Method, consisting of 140 interview items.

As prerequisite to the use of these scales in studies, the interview should be conducted only by psychiatrists who have intensively trained on more than ten assessment exercises for several days. These are designed to further derive 38 syndromes, 50 subclasses, as well as Catego classes by assessing 140 symptoms. From the 140 symptoms, comprising 38 syndromes listed in Table 5.10, another 22 potential symptoms still remain excluded. Catego classes obtained by computer deconvolution of these 38 syndromes can roughly be divided into classes S (Schizophrenia), P (Paranoia), M (Mania), D (Depression), R (Restrained depression), N (Neurosis), A (Anxiety), and O (Others). Some of them can be reclassified into two stages of + or ? based on their respective accuracy levels. Results from preceding studies have revealed a high correlation between Catego class and clinical diagnosis. This indicates that Catego class S+ shown in Table 5.8 should be considered to agree with *narrowly defined schizophrenia*.

Table 5.10 Thirty-eight syndromes derived from symptoms of PSE

Syndromes	The number of each symptom and the descriptions constituting the syndrome
1. (NS) Nuclear syndrome	55. Thought intrusion, 57. Thought commentary, 58. Thought withdrawal, 62. Voices about patient, 71. Delusions of control, 81. Delusions of alien penetration, 82. Primary delusions
2. (CS) Catatonic syndrome	116. Mannerisms and posturing, 119. Catalonic movements
3. (IS) Incoherent speech	135. Neologisms, 136. Incoherence of speech
4. (RS) Residual syndrome	60. (2) Hears muttering, whispering, 118. Behaves as if hallucinated, 132. Nonsocial speech
5. (DD) Depressive delusions and hallucinations	61. (2) Depressive hallucinations, 88. Delusions of guilt, 91. Hypochondriacal delusions (bowels blocked up), 92. Delusions of catastrophe
6. (SD) Simple depression	19. Inefficient thinking, 23. Depressed mood, 24. Hopelessness, 25. Suicidal plans or acts, 121. Depression on examination
7. (ON) Obsessional neurosis	44. Checking and repeating, 45. Cleanliness and rituals, 46. Obsessional ideas and rumination
8. (GA) General anxiety	11. Anxiety, 14. Panic attacks, 120. Anxiety on examination
9. (SA) Situational anxiety	15. Situational anxiety, 17. Specific phobias, 18. Anxiety avoidance
10. (HT) Hysteria	64. (2) Dissociative hallucinations (not subcultural), 100. Dissociative states, 101. Conversion symptoms, 122. Histrionic
11. (AF) Affective flattening	128. Blunted affect
12. (HM) Hypomania	41. Subjective euphoria, 42. Ideomotor pressure, 43. Grandiose ideas and actions, 123. Hypomanic affect, 137. Hypomanic content of speech
13. (AH) Auditory hallucinations	63. Voices to patient (not depressive)
14. (PE) Delusions of persecution	74. Delusions of persecution
15. (RE) Delusions of reference	72. Delusions of reference, 73. Delusions of misinterpretation
16. (GR) Grandiose and religious delusions	76. Delusions of grandiose ability, 77. Delusions of grandiose identity, 78. Religious delusions
17. (SF) Sexual and fantastic delusions	59. Thoughts read, 70. (2) Delusional elaboration of hallucinations, 75. Delusions of assistance, 79. Delusional explanation (hypnotism, etc.), 80. Delusional explanation (rays, etc.), 84. Morbid jealousy, 85. Delusions of pregnancy, 86. Sexual delusions, 87. Fantastic delusions, 89. Delusions concerning appearance, 90. Delusions concerning lack of organs
18. (VH) Visual hallucinations	66. (2) Visual hallucinations
19. (OH) Olfactory hallucinations	68. Olfactory hallucinations, 69. Delusions that patient smells
20. (OV) Overactivity	112. Gross excitement, 113. Irreverent behavior, 115. Embarrassing behavior
21. (SL) Slowness	110. Slowness and underactivity, 130. Slow speech, 133. Muteness, 134. Restriction of quantity of speech

(continued)

Table 5.10 (continued)

Syndromes	The number of each symptom and the descriptions constituting the syndrome
22. (NP) Nonspecific psychosis	49. Unfamiliarity and delusional mood, 50. Heightened perception, 52. Changed perception, 53. Changed perception of time, 60. (1) Hears music, tapping, etc., 61. (1) Hears voice calling name, 66. (1) Minor visual hallucinations, 70. (1) Other minor hallucinations, 94. Evasiveness concerning delusions, 102. Clouding or stupor, 109. Bizarre appearance, 117. Stereotypies, 125. Suspicion, 126. Perplexity, 129. Incongruous affect
23. (DE) Depersonalization	47. Derealization, 48. Depersonalization
24. (ED) Special features of depression	29. Self-depreciation, 32. Guilty ideas of reference, 33. Guilt, 51. Dulled perception, 54. Lost affect
25. (AG) Agitation	111. Agitation on examination
26. (NG) Self-neglect	108. Self-neglect
27. (IR) Ideas of reference	31. Ideas of reference
28. (TE) Tension	5. Tension pains, 7. Muscular tension, 8. Restlessness
29. (LE) Lack of energy	36. Subjective anergia
30. (WO) Worrying, etc.	4. Worrying, 6. Tiredness, 10. Nervous tension, 21. Neglect through brooding, 35. Delayed sleep
31. (IT) Irritability	40. Irritability, 124. Hostile irritability
32. (SU) Social unease	16. Anxiety on meeting people, 28. Social withdrawal, 30. Lack of self confidence
33. (IC) Loss of interest and /or concentration	20. Poor concentration, 22.Loss of interest
34. (HY) Hypochondriasis	9. Hypochondriasis
35. (OD) Other symptoms of depression	27. Morning depression, 34. Loss of appetite, 37. Early waking, 38. Loss of libido, 39. Premenstrual exacerbation
36. (OR) Organic impairment	67. Delirious visual hallucinations, 103. Organic impairment of memory
37. (SC) "Subcultural" delusions or hallucinations	64. (1) "Subcultural" hallucinations, 83. "Subcultural" delusions
38. (DI) Doubtful interview	139. Misleading answers, 140. Interview doubtfully adequate

Subsequently, four subscores can be derived from the computer program in order to grasp clinical mental states. All 36 syndromes that serve as baseline of the subscores, excluding organic disorder (OD) and doubtful interviews (DI), are listed in Table 5.11. These subscores appear to be extrapolated from the clinical characteristics of each case into general psychiatric terms, and to that extent it is useful to understand the pathological conditions, not on a disorder basis.

It is often said that assessment results of clinical symptoms, such as experiential or positive symptoms, roughly agree, whereas those based on expressive symptoms, such as facial expression, attitudes, or behaviors, are prone to disagreement. Table 5.12 compares the results of assessment with subscores in the Nagasaki

Table 5.11 PSE subscores

DAH: delusions and hallucinations

(NS) Core syndrome, (DD) depressive delusion and hallucination, (AH) auditory hallucinations, (PE) delusion of persecution, (RE) delusion of reference, (GR) grandiose and religious delusions, (SF) sexual and fantastic delusions, (VH) visual hallucinations, (OH) olfactory hallucinations, (SC) "subcultural" delusions or hallucinations

BSO: behavior, speech and other

(CS) Catatonic syndrome, (IS) incoherent speech, (RS) residual syndrome, (AF) affective flattening, (HM) hypomania, (OV) overactivity, (SL) slowness, (NP) nonspecific psychosis, (AG) agitation, (NG) self-neglect

SNR: specific neurotic symptoms

(SD) Simple depression, (ON) obsessional neurosis, (GA) general anxiety, (SA) situational anxiety, (HT) hysteria, (ED) special features of depression

NSN: nonspecific neurotic symptoms

(DE) Depersonalization, (IR) ideas of reference, (TE) tension, (LE) lack of energy, (WO) worrying, etc., (IT) irritability, (SU) social unease, (IC) loss of interest and /or concentration, (HY) hypochondriasis, (OD) other symptoms of depression

Table 5.12 Reliability among assessors in PSE subscores (based on data of DOSMeD Nagasaki, 1983)

PSE subscore	Mean score ($n=25$)			Total score $A+B$	Score bias among assesors	Bias (%)
	A	B	t			
DAH: delusion and hallucination	8.2	8.0	0.102	405	33	8.15
BSO: behaviors, spoken dialogues, and others	7.0	5.5	1.034	314	53	16.88*
SNR: specific neurosis syndrome	4.9	4.4	0.462	232	28	12.07
NSN: nonspecific neurosis syndrome	9.2	9.1	0.049	457	30	6.57
Total (PSE score)	29.3	27.0	0.468	1408	118	8.38

*$p<0.005$

DOSMeD study sample. This implies that, between provisionally named assessors A and B, the gap in their assessment results between *delusion and hallucination* (*DAH*) and *nonspecific neurosis* (*NSN*), which typically involve positive symptoms, such as hallucination and delusion, was still tolerable. A great gap was, however, seen in results between *specific neurosis* (*SNR*) and *behaviors, spoken dialogues, and others* (*BSO*). As mentioned, particularly in BSO, a significant difference can be identified between the two assessors, suggesting meticulous attention to psychotic symptoms during the assessment process.

5.4 Help-Seeking Patterns and Recognition by Family Members (Key Persons) in New Schizophrenia Cases in Nagasaki

The treatment patterns (help-seeking patterns) of newly onset schizophrenia cases meeting the inclusion criteria for this study have provided intriguing information that is relatively new to us. Table 5.13 lists the first contact institutions where new schizophrenia outpatients and their families sought consultation. While 60 of all 107 subjects (56 %) had the first consultation at mental hospitals and other providers of mental health services, the rest varied from case to case, and, particularly, more than 10 % turned to religious workers and/or traditional faith healer, or police officers. This phenomenon may, however, reflect the geographical characteristics of Nagasaki and is not thus intended to be seen as general tendency immediately. However, this may remind providers of medical services that psychiatrists are not always the first contact for consultation, even for schizophrenia patients.

As a treatment pattern in most cases, family members of a patient would visit different institutions along with the patient. But in some cases, family members would seek help on their own in consideration of the patient's condition. In this study, family members of 32 subjects sought advice alone without patients, while family members of 24 subjects contacted a psychiatrist for their initial consultation along with patients, and those of 51 subjects resorted to some other experts before the initial psychiatric consultation. Following this, we compared the treatment patterns to the recognition patterns by family members of the problems patients have. The results suggested that families that resorted to some other expert for consultation without patients came to the earliest recognition of association of patients' problems or symptoms with any psychiatric disorders, and showed the shortest interval of 4.87 weeks from the operationally determined date of onset until the first contact with psychiatrists along with patients among any other pattern groups. Conversely, first consultation along with patients was found to lead them to a difficulty in properly understanding the information given there, and this confusion deferred them from coming into contact with psychiatrists, resulting in a delay (compared with a 9.62-week interval in the group who contacted a psychiatrist along with patients and a 10.66-week interval in the group who contacted a nonpsychiatric expert along with patients) (see Table 5.14).

Table 5.13 The supporting institutions for the initial consultation of the patient taken by the family

Institutions for consultation	Cases (%)
Psychiatrist/other psychiatric expert	60 (56.1)
General physician/other medical expert (nonpsychiatry)	27 (25.2)
Nurse/social worker/other health-care worker	7 (6.5)
Police officer	5 (4.7)
Traditional healer (e.g., medicine man)	2 (1.9)
Clergyperson/spiritual advisor	6 (5.6)
Total	*107*

Table 5.14 Three help-seeking patterns and recognition by family members in 107 schizophrenia cases (%)

	A group (51 cases)	B group (32 cases)	C group (24 cases)
	Joint help-seeking by patient and family members	Help-seeking by family members	Seeking psychiatric consultation from the beginning
	The patient and the family visited other medical institution before psychiatric consultation	Only the family initially visited some supporting intuition	The patient and the family visited a psychiatric hospital for initial consultation
Symptoms of psychiatric disorder	25 (49.0)	22 (68.8)	8 (33.3)
Other problems	26 (51.0)	10 (31.3)	16 (66.7)
Time from onset to initial psychiatric consultation	10.66 weeks	4.87 weeks	9.62 weeks

Regarding which factors gave rise to seeking consultation and treatment, recognition patterns by family members of problems patients have still remains unknown. One possible factor may presumably be considered to be the significance ascribed to the patient's behavioral problems based on positive symptoms, and, in this DOSMeD study, key family members were interviewed additionally to identify the recognition patterns of *early symptoms* expressed by the patients to them.

Based on the PPHS scale, 25 early symptoms were culled with reference to prodromal symptoms exemplified in the DSM system and the like, and then translated into daily terms for interview with *key persons* by researchers. Once such symptoms were responded to positively, additional dialogue interview with the key person was carried out in an attempt to understand exactly the relevant symptoms, including the timing of expression. Separately from the self-reported timing of expression, the researcher (psychiatrist) estimated the exact timing and pattern of onset according to his accumulated clinical experience and all pertinent information.

Out of 107 subjects, sufficient information was provided by key persons for 93 subjects. The information collected was classified roughly into three categories in accordance with the schizophrenia subtype specified in the ninth edition of the International Classification of Diseases (ICD-9) and are compared in Table 5.15, which lists all the symptoms in descending order of frequency.

As this table implies, recognition pattern of behavioral abnormalities in patients by key persons varies according to the subtype of schizophrenia. On average, patients responded positively to 5.9 early manifestations. With a symptom-focused view in the left-most column of the table (titled in the table as *all cases*), the statement *The patient became noticeably inactive on the job and at home* was most common with 69.9 %, followed by *Avoidance or refusal of interaction with others* and *Reduction in appetite, sleep, sexual desire,* etc. with 51.6 % each. Least observed symptoms were *Temporary memory loss*, *Talking about someone already passed away as if he or she is still alive*, and *Talking about an unimaginably ambitious plan* with 3.2 % each. Most symptoms ranking in the top 10 involved abnormal behaviors or facial expression. Viewing these by schizophrenia subtypes, such as the hebephrenic or paranoid types, no significant difference was observed in hebephrenic schizophrenia among all categories, but symptoms exhibited by paranoid schizophrenia patients not only increased to 6.9 items on average but also apparently reflected phenomena characteristic of the disease, including *Showing inexplicable (delusional) ideas* as the top symptom. There was a substantial difference from the overall results of those seen in hebephrenic cases.

Subsequently, a similar analysis was performed in 51 acute-onset cases in which florid symptoms of schizophrenia (refer to SS were exhibited within one month from the expression of early manifestations), and whether or not such symptoms gave rise to first consultation with psychiatrists was examined. Subtypes of the sample subjected to this analysis include 19 hebephrenic cases, 17 paranoid cases, 8 acute schizotypal episodic cases, and others. Table 5.16 ranks the top 13 of 25 items in the early symptoms list responded to positively by interviewees. In comparison with the results in the 93 cases shown earlier, no significant difference was observed in expression frequency or ranking of symptoms among the top 1–5. As expected,

Table 5.15 Early symptoms recognized by the key person in the patients of the DOSMeD study

	All cases	Hebephrenic type	Paranoid type	Others
Marked decrease in activities at the office or home	69.9	81.2*	57.1*	58.8
Avoidance or refusal of interaction with others	51.6	62.5*	39.3*	41.2
Decreased sleep, appetite or sexual desire	51.6	41.7	57.1	70.6
Difficult to hear and understand what he/she says	41.9	37.5	57.1	29.4
Inexplicable claim (delusional content)	40.9*	31.3	67.9*,	23.5
Behaviors suspecting auditory hallucinations	37.6	39.6	46.4	17.6
Causeless unrest and iracibility over several days	34.4	29.2	46.4	29.4
Words as if he/she was harmed (delusional content)	32.3*	22.9	53.6*,	23.5
He/she was surprised or worried for several days without specific reason	32.3	35.4	32.1	23.5
He/she looked very sad, overcome with sorrow and lost hope	31.2	35.4	28.6	23.5
He/she embarrassed others with rude manner	25.8	18.8	32.1	35.3
Decreased interest in self-appearance	22.6	29.2	14.3	17.6
Complaints of continuous physical pain or queer feeling	20.4	18.8	28.6	11.8
Sudden move to another place without any reason	19.4	27.1	14.3	5.9
Self-injury or attempted suicide	17.2	16.7	10.7	29.4
Excitement continuing for several days	10.8	10.4	14.3	5.9
He/she destroyed things or set a fire	10.8	12.5	14.3	0
He/she caused physical violence to others	9.7	6.3	17.9	5.9
Lack of consideration in behavior	8.6	4.2**	17.9**	5.9
He/she believed that he/she had a fatal physical disease	5.4	4.2	10.7	0
He/she got spiritual more than before	4.3	2.1	7.1	5.9
He/she spent recklessly and unreasonably more than before	4.3	6.3	3.6	0
He/she complained of a temporary loss of memory	3.2	6.3	0	0
He/she spoke as if some dead person were still alive	3.2	0	10.7	0
He/she spoke about an unimaginably ambitious plan	3.2	2.1	7.1	0
The number of subjects	93	48	28	17

There is no significant difference between the hebephrenic type and all cases

Chi-squared test: *$p<0.05$ (comparison between the paranoid type and the hebephrenic type and between the paranoid type and all cases). $p<0.05$ (comparison between the paranoid type and the hebephrenic type). **$p<0.1$ (comparison between the paranoid type and the hebephrenic type)

Table 5.16 Correlations between consultation and early symptoms perceived by key persons in patients presenting acute schizophrenia

	Rate of affirmation	Correlation with consultation
Marked decrease in activities at the office or home	68.6	Negative correlation*
Decreased appetite, sleep, or sexual desire	62.7	
Inexplicable claim (delusional content)	45.1	Positive correlation*
Difficult to hear and understand what he/she says	41.2	Positive correlation*
Avoidance or refusal of interaction with others	39.2	Negative correlation*
He/she was surprised or worried for several days without specific reason	39.2	
Words as if he/she was harmed (delusional content)	37.3	
Causeless unrest and iracibility for several days	35.3	
He/she embarrassed others with rude manner	31.4	
He/she looked very sad, overcome with sorrow and lost hope	31.4	Negative correlation*
Behaviors suspecting auditory hallucinations	29.4	Positive correlation*
Sudden move to another place without any reason	17.6	Positive correlation*
He/she caused physical violence to others	5.9	Negative correlation*

*$p<0.1$–0.01

the top item, *The patient became noticeably inactive on the job and at home* was found to negatively correlate with early consultation; the first cases with positive correlation with early consultation had the third and fourth symptoms.

Not surprisingly, a summary of the results may suggest that the effect of positive symptoms as a cause leading to consultation along with patients is greater than that of negative symptoms.

In brief, I would also like to mention the duration of untreated psychosis (DUP), which has recently been a controversial topic. Mean DUP scores were 8.2 ± 11.39 months in males ($n=54$), 12.02 ± 22.3 months in females ($n=43$), and 9.9 ± 17.23 months in both sexes combined ($n=97$) (median 4.0 months each). This may suggest a slight delay in treatment start for female patients.

We will touch on DUP again later. Its short- and long-term effects on patients' outcomes have also been pointed out.

5.5 Clinical Psychiatric Characteristics of New Schizophrenia in Nagasaki

5.5.1 Subtypes of Schizophrenia

Table 5.17 lists subtypes of schizophrenia included in the DOSMeD study categorized according to the ICD-9 classifications used in the initial assessment. As many as 53 of 107 subjects, about half of the sample, were categorized into hebephrenic

Table 5.17 Subtypes identified at initial assessment in new patients with schizophrenia

ICD-9 codes and subtypes	Male	Female	Total (%)
295.0 Schizophrenia simplex	1	1	2 (1.9)
295.1 Disorganized type schizophrenia	29	24	53 (48.5)
295.2 Catatonic schizophrenia	0	2	2 (1.9)
295.3 Paranoid type schizophrenia	15	16	31 (29.0)
295.4 Schizophreniform disorder	4	1	5 (4.7)
295.5 Latent schizophrenia	4	0	4 (3.7)
295.6 Residual type	0	1	1 (0.9)
295.7 Schizoaffective disorder	1	0	1 (0.9)
295.8 Other specified types of schizophrenia	2	0	2 (1.9)
295.9 Unspecified schizophrenia	5	1	6 (5.6)
Total	*61*	*46*	*107 (100.0)*

schizophrenia (ICD-9/295.1). This predominant position of hebephrenic schizophrenia in subtype distribution is considered extremely common in Japan; however, this pattern was found to be unique to Japan and seldom occurred in other DOSMeD study participating countries.

As Table 5.18 suggests, in most foreign countries, the hebephrenic type was shown to be rare, while the paranoid type was recognized as major. For example, among all schizophrenia subtypes, hebephrenic schizophrenia accounted for merely 5.8 % and 4.3 % in developed countries other than Japan and in developing countries, while paranoid schizophrenia accounted for 34.4 % and 22.9 %, respectively, Japan appeared to have rather a similar expression frequency to that in developing countries. On the other hand, a considerably high expression frequency of the catatonic type (ICD-9/295.2) and acute schizotypal episodes (ICD-9/295.4) was characteristic of developing countries.

Coinciding with the completion of recruitment for the DOSMeD study sample, the ICD-9 was revised to ICD-10. For a long-term follow-up of the clinical course and outcomes of the subjects, the subjects were categorized anew into subtypes as established in the ICD-10. Since ICD-9 had no distinct standardized criteria for diagnosis, schizophrenia was usually categorized into subtypes according to the conventional diagnostic procedures, which vary according to cultures. Categorization based on a common concept was not possible unless using PSE or other scales. On the other hand, ICD-10 Chap. V (F) enables a categorization of subtypes based on the disease-based standardized Clinical Description and Diagnostic Guideline (CDDG) and Diagnostic Criteria for Research (DCR) established after the DSM system, as summarized in Table 5.19.

As shown in this table, 33 of 107 (30.8 %) subjects were re-categorized into paranoid-type schizophrenia, including 14 additional ICD-9 hebephrenic cases, which ranked number 1 also in Japan (Nagasaki). In comparison with the distribution in Table 5.18, an almost similar pattern was obtained to those in foreign countries.

Table 5.18 Subtype distribution in 295 of ICD-9 (%)

	295.0	295.1	295.2	295.3	295.4	295.5	295.6	295.7	295.8	295.9
All countries (n=1,151)	2.3	9.2	5.2	28.8	23.8	3.7	2.3	5.9	14.2	4.4
Developing countries (n=506)	1.6	4.3	10.3	22.9	40.3	1.8	4.0	3.2	4.3	7.3
Developed countries (n=645)	2.9	13.0	1.2	33.5	10.9	5.3	0.9	8.1	22.0	2.2
Excluding Japan (n=538)	3.2	5.8	1.1	34.4	12.1	5.6	0.9	9.5	26.0	1.5
Nagasaki (n=107)	1.9	48.5	1.9	29.0	4.7	3.7	0.9	0.9	1.9	5.6

295.0 Schizophrenia simplex; 295.1 disorganized type schizophrenia; 295.2 catatonic schizophrenia; 295.3 paranoid type schizophrenia; 295.4 schizophreniform disorder; 295.5 latent schizophrenia; 295.6 residual type; 295.7 schizoaffective disorder; 295.8 other specified types of schizophrenia; 295.9 unspecified schizophrenia

Table 5.19 Reassessment with ICD-10 in schizophrenic patients categorized under ICD-9

ICD-10F/ICD-9	295.1 Disorganized type	295.3 Paranoid type	Others	Total (%)
F20.0 Paranoid schizophrenia	14	18	1	33 (30.8)
F20.1 Hebephrenic schizophrenia	15	0	1	16 (15.0)
F20.3 Undifferentiated schizophrenia	9	4	3	16 (15.0)
F23 Acute and transient psychotic disorders	7	6	7	20 (18.7)
F25 Schizoaffective disorders	0	0	1	1 (0.9)
Others	8	3	10	21 (19.6)
Total	*53*	*31*	*23*	*107 (100.0)*

Table 5.20 International comparison of patients with schizophrenia reassessed under ICD-10F (%)

ICD-10F	Nagasaki	Nottingham	Prague	Chandigarh/u
F20.0 Paranoid type schizophrenia	30.8	64.3	48.7	26.8
F20.1 Hebephrenic type schizophrenia	15.0	17.1	2.6	0.9
F20.3 Undifferentiated schizophrenia	15.0	0	0	0
F23 Acute and transient psychotic disorders	18.7	5.7	9.2	42.0
F25 Schizoaffective disorders	0.9	7.1	21.1	0
Others	19.6	5.8	18.4	30.3
Total (n)	*107*	*70*	*76*	*112*

Note: The data excluding that of Nagasaki were recalculated based on "recovery from schizophrenia" (2007) [16].

If we now compare the ICD-10/F data from other centers retrieved from the book that summarizes the follow-up study results published in 2007 (see Table 5.20), the small number of paranoid schizophrenia patients is still obvious in Japan among other developed countries. Expression frequency of acute transient psychiatric disorder (F23) is exceptionally higher than any other industrial countries, except for the high rate seen in Chandigarh, India. Only 15 of 53 ICD-9 hebephrenic cases remained the same subtype also in the ICD-10/F categorization, and they account for merely 15 % of all cases.

Such a unique result would imply that the hebephrenic type has been the focus in psychiatric education and clinical practice as the major subtype of schizophrenia in Japan. When compared to global trends, however, there should have been without doubt a slight gap in categorization of subtypes between Japan and foreign countries. In particular, through the application of the stringent diagnostic criteria of the ICD-10/F to each clinical case, the distribution pattern of subtypes in Japan became comparable to those in other industrial countries. If the categorization based on the standardized diagnostic criteria has, in fact, brought the distribution pattern of subtypes in Japan close to the global one, conventional categorization of diseases, in particular subtypes, definitely needed to be overhauled.

5.5.2 Psychiatric Symptoms Schizophrenia

In the DOSMeD study, psychotic symptoms were assessed using the PSE. In the assessment of the clinical manifestation of psychiatric disorders, the PSE is often criticized for assessment results that are prone to disagreement, resulting in inappropriate or inconsistent diagnosis. Factors associated with reliability of assessment and diagnosis include: (1) subjective variance, (2) occasion variance, (3) criterion variance, (4) information variance, and (5) observation variance.

It is widely known that eliminating such variances is an urgent issue to be addressed to reinforce the process of psychiatric clinical studies. When serving as an information source, we may have to take this point a little bit more into consideration. The APA focused on (3) and (5) during the development process of the DSM system, proposing diagnostic criteria and establishing a structured interview schedule for symptom assessment. Meanwhile, WHO has proposed some symptom assessment scales during the development process of the ICD system, using a similar procedure to that of the APA. It is not fully recognized in Japan yet, but the WHO Collaborating Center in Nagasaki has also been leveraging the above two approaches in our assessment process with an intense use of the PSE. The following paragraphs explain about psychotic symptoms in the first episode of schizophrenia in DOSMeD subjects in Japan obtained using the PSE.

It is pointed out that clinical manifestation has recently changed in various psychiatric disorders. For example, it is controversially noted that most psychiatric disorders, including schizophrenia, seem to have been lessening in severity. However, psychotic symptoms have not always been assessed systematically, so there may be no evidence to immediately conclude that symptoms have been particularly stereotyped, lessening in severity, or increasingly exhibiting neurotic aspects. I myself certainly cannot deny my impression that schizophrenia symptoms around the time when I became a psychiatrist presented with significant agitation in addition to various positive symptoms. Looking back at past profiles, there might have been some variation in symptoms. It was positive symptoms that were mainly under intense scrutiny in the routine diagnostic process, and rand-symptoms, such as negative symptoms, were possibly apt to be ignored as being ancillary. Neuropsychiatric symptoms might have already been observed in schizophrenia patients at that time, but without any interview schedule established for a comprehensive assessment as we have today, sufficient attention was not given to them.

Next, we will take a look at the profile of PSE symptom categories observed in 98 of 107 first-episode schizophrenia subjects in the DOSMeD study at the initial consultation. This profile is shown in a standard layout proposed in a global study on schizophrenia as the main study subject using PSE, and symptoms are not exactly listed in order of the category number, consisting largely of subscore groups such as BSO, DAH, and SNR, as seen in Table 5.21.

As seen in Fig. 5.3, new schizophrenia that developed about 25 years ago had already presented with various symptoms, such as depression and neurotic symptoms, in addition to positive and major symptoms characteristic of this disorder. An unexpected outcome was obtained reflecting the distribution of clinical subtypes

Table 5.21 Syndromes used in distribution chart

Syndromes	PSE subscores
(OV) Overactivity, (SL) slowness, (CS) catatonic syndrome, (IS) incoherent speech, (RS) residual syndrome, (AF) affective flattening	BSO: behaviors, spoken dialogues, and others
(NS) Nuclear syndrome, (AH) auditory hallucinations , (PE) delusions of persecution, (RE) delusions of reference, (VH) visual hallucinations, (OH) olfactory hallucinations, (GR) grandiose and religious delusions	DAH: delusion and hallucination
(HM) Hypomania	BSO: behaviors, spoken dialogues, and others
(DD) Depressive delusions and hallucinations	DAH: delusion and hallucination
(ED) Special features of depression, (SD) simple depression, (GA) general anxiety, (SA) situational anxiety, (ON) obsessional neurosis	SNR: specific neurosis syndrome

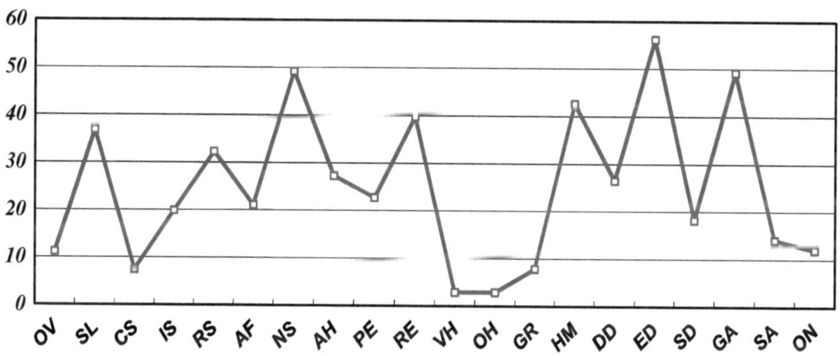

Fig. 5.3 The profile of symptoms in new patients with schizophrenia (at the initial consultation). *OV* overactivity, *SL* slowness, *CS* catatonic syndrome, *IS* incoherent speech, *RS* residual syndrome, *AF* affective flattening, *NS* nuclear syndrome, *AH* auditory hallucinations, *PE* delusions of persecution, *RE* delusions of reference, *VH* visual hallucinations, *OH* olfactory hallucinations, *GR* grandiose and religious delusions, *HM* hypomania, *DD* depressive delusions and hallucinations, *ED* special features of depression, *SD* simple depression, *GA* general anxiety, *SA* situational anxiety, *ON* obsessional neurosis

of schizophrenia in that expression frequency of catatonic symptoms was extremely low, visual and olfactory hallucinations were considerably rarely observed among any other hallucinatory episodes, and, instead, a substantially high rate was shown of depressive and anxiety symptoms. This may be considered as one piece of evidence that implies that the global and comprehensive understanding of psychiatric symptoms that patients have was achievable only through assessment based on a

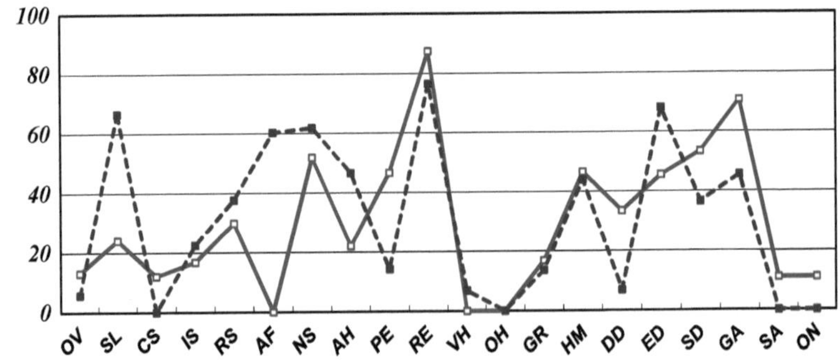

Fig. 5.4 The profile of symptoms in new patients with schizophrenia (at the initial consultation) by onset patterns (*solid line*: acute onset case, $n=17$; *dotted line*: latent onset case, $n=13$)

structured interview schedule. This figure was created by compiling schizophrenia symptom distributions by subtypes and by onset patterns into one, Fig. 5.4 shows a distribution of 30 cases that were available for analysis of onset pattern.

Three onset patterns analyzed in this study included acute, subacute, and chronic. Acute onset is operationally defined as *remarkable psychotic state progressing from a nonpsychotic state within a few days (up to one week) without any prodromal symptoms*; subacute onset as *obvious psychotic state progressing from symptom expression within one month*; and chronic onset as *psychotic state gradually progressing over one month*. As estimated, Fig. 5.4 suggests that anxiety and depressive symptom were more significant in acute onset cases; latent onset cases were more apparently prone to inactivity and apathia.

As part of an attempt to demonstrate the achievement of the comprehensive symptom assessment instrument using PSE, I compared the results from preceding studies on Schneider's First Rank Symptoms (FRS); Symptome ersten Ranges. First-rank symptoms proposed by Schneider [17] are symptom categories that champion diagnoses of schizophrenia based on the form, if the irrelevance of exhibited symptoms to any physical diseases is confirmed. Although the terms are often used descriptively by mental health professionals in Japan, the strict positive ratio and the reliability of using FRS for the diagnosis of schizophrenia have been questioned. In addition, symptoms listed as FRS appear to have never been strictly validated. We have rigorously extracted PSE symptoms associated with FRS and compared them to the findings of foreign reports.

Table 5.22 lists eight items of FRS that are cross-referable with the endpoints of PSE. Hirohiko Harima has indicated in his recent translation *Kurt Schneider, new edition of Clinical Psychopathology, the 15th edition of the original work* (Tokyo: Bunko-do, 2007) [18] that Schneiderian FRS consist of seven psychotic symptoms.

Table 5.23 shows the assessment results of eight symptoms in accordance with the PSE listed in Table 5.22.

Table 5.24 indicates the positive rates of FRS, suggesting that hebephrenic schizophrenia obviously distinguishes from other subtypes with its highest

Table 5.22 Schneider's first rank symptoms

• Gedankenlautwerden; audible thought	PSE-56(1)
• Auditory hallucinations (Hören von Stimmen in der Form von Rede und Gegenrede)	PSE-62(2)
• Hören von Stimmen, die das eigene Tun mit Bemerkungen begleiten; voices arguing, running commentary on patient's actions	PSE-62(1)
• Gedankenentzug; thought withdrawal andere Gedankenbeeinflussungen, Gedankeneingebung; thought insertion	PSE-58(2)
• Gedankenausbreitung; thought broadcasting or believe others to think the same thoughts	PSE-55(1,2)
• Whanwaurnehmung; delusional perception	PSE-56(2)
• Alles von anderen Gemachte und Beeinflusste auf dem Gebiet Fuhlens, Strenbens (der Trieb) und des Wollens; feelings, impulses, volitional acts thought to be due to others leibliche Beeinflussungserlebnis; somatic passivity: experiences of physical interference by outside influences	PSE-82(1,2) PSE-71(1,2)

Table 5.23 Positive rates of respective symptoms (%, values in parentheses show full delusions)

	Hebephrenic type (n=48)	Paranoid type (n=28)	All (n=93)
Audition of thoughts	2.0	–	2.2
Auditory hallucinations, spoken to the subject	14.6	25.0	19.4
Auditory hallucinations, discussed about the subject	8.3	14.3	10.8
Thought withdrawal	–	3.6	2.2
Thought insertion	16.6 (8.3)	14.3 (7.1)	18.3 (7.5)
Thought broadcasting	16.6	25.0	19.4
Delusion of being controlled	20.8 (8.3)	35.7 (21.4)	25.8(14.0)
Primary delusion (delusional perception)	35.3 (22.9)	50.0 (39.3)	37.6(26.9)

Table 5.24 Positive rates in Schneider's FRS

	Hebephrenic type (n=48)	Paranoid type (n=28)	Others (n=17)	All subjects (n=93)
0	24 (50)	9 (32)	7 (41)	40 (43)
1 and up	24 (50)	19 (68)	10 (59)	53 (57)
2 and up	13 (27)	12 (43)	6 (35)	31 (33)
3 and up	7 (15)	7 (25)	4 (24)	18 (19)
4 and up	3 (6)	2 (7)	1 (6)	6 (7)
5 and up	–	–	1 (6)	1 (1)
6 and up	–	–	–	–

Percentage is provided in parentheses

Table 5.25 Positive rate of Schneider's FRS in patients clinically diagnosed with schizophrenia and without schizophrenia [19]

Schizophrenia	No.	Areas	Positive rate (%)
Prospective study			
Mellor (1970)	166	UK	72
Carpenter et al. (1973)	103	USA	53
Carpenter et al. (1974)	801	IPSS	57
Strauss et al. (1977)	57	USA	43
Silverstein et al. (1978)	72	USA	24
Bland et al. (1978)	92	Canada	53
Zarrouk (1978)	92	Saudi-Arabia	56
Preiser et al. (1979)	52	USA	48
Nakane et al. (1982)	93	DOSMeD-Japan	57
Case record			
Huber (1967)	195	Germany	72
Taylor (1972)	78	USA	28
Abrams et al. (1973)	71	USA	34
Koehler et al. (1977)	210	Germany	33
Other diseases	*n*	Clinical diagnosis	Positive rate (%)
Taylor et al. (1973)	52	Mania	11.5
Carpenter et al. (1973)	39	Mania	23
Carpenter et al. (1974)	66	Mania	23
	119	Depression	16
	123	Neurotic/personality disorder	4
Taylor et al. (1975)	53	Mania	8
Silverstein et al.(1978)	54	Nonschizophrenic disorder	4
Preiser et al. (1979)	36	Non-schizophrenic disorder	14

frequency in which no FRS is applicable, and some FRS were exhibited by 53 % of first-episode schizophrenia patients that were sample subjects included in the DOSMeD study.

Table 5.25 compares the above results to the findings of the preceding studies reported in Europe and the USA. A positive FRS ratio varies according to the study method, though, ranging from 24 % to 72 % in prospective studies and 28 % to 72 % in case records, showing a notably wide gap between the lowest and highest ratios but a similar variance overall. The lower half of the table indicates a positive ratio of FRS in non-schizophrenia cases, with an overall definitely lower positive ratio than that of schizophrenia cases. No additional data for this table are available in Japan.

Another topic has caught my attention in the course of the FRS study. FRS has a symptom bracket categorized as *Audible thoughts* which had been interpreted as *hallucination of a dialogue between the patient and hallucinatory voices* for a long time, ever since it was introduced in Japan. During the translation process of PSE into Japanese, however, this was found to be distinguished from *Voices heard arguing*.

Table 5.26 Frequencies (%) of audible thoughts in schizophrenic patients observed in DOSMeD/Nagasaki

	63. Spoken-to-the subject			62. Discussed-about-the subject		
	M	F	M+F	M	F	M+F
Total (*n*=93)	12.2	22.8	17.2	20.4	18.2	19.4
Hebephrenic type (*n*=48)	9.1	34.6	22.9	18.2	11.5	14.6
Paranoid type (*n*=28)	13.3	7.7	10.7	20.0	20.8	25.0

The latter had been regarded as a symptom intrinsically characteristic of schizophrenia that should rather belong to FRS. On the contrary, *Dialogue with hallucinatory voices* is deemed to correspond to *Hysterical syndrome* (*HT*) in 38 PSE syndromes with the recognition as *dissociative hallucination*. Table 5.26 compares the assessment and identification process of the above two types of *hallucination of a dialogue* in DOSMeD subjects in Nagasaki.

Of all the subjects, 17.2 % experienced hallucination of *dialogue between the patient and hallucinatory voices* and 19.4 % *hallucinatory voices engaged in debate or argument about the patient*. Analyzing it further by subtypes, *hallucinatory voices engaged in debate or argument about the patient* was most frequently experienced by paranoid schizophrenia subjects, whereas *dialogue between the patient and hallucinatory voices* was experienced by hebephrenic subjects. In particular, proliferation of *dialogue between the patient and hallucinatory voices* in female hebephrenic subjects conclusively balanced the overall frequency of both hallucinations of a dialogue. In accordance with these findings, *hallucinatory voices engaged in debate or argument about the patient* may be appropriate to be considered as a significant FRS for schizophrenia.

5.5.3 Life Events and the Development of Schizophrenia

While quite a large number of study reports on depression-inducing life events are available, no consistent view on life events associated with schizophrenia has been achieved, even though schizophrenia has its own vulnerability-stress-coping model, and *life events* is one of the constituent elements of this model. Life events thus became a pivotal endpoint in the DOSMeD study.

To start with the establishment of a life event assessment instrument, some existing instruments used in preceding studies, including the Social Readjustment Rating Scale (SRRS) of Holmes and Rahe, the Life Event Checklist of Brown and Birley, and the List of Life Events of Paykel et al. were weighed, and a new WHO-specific LES was produced. Based on this, we translated the complete original English text into Japanese and composed and published a guide as private edition.

Based on the definition of life event as *an objective experience at a significant level that forces most individuals who experience such life events to change their usual life* [20], the *Life Event List* was drafted in consideration of cultural diversity overall as well as of the DOSMeD study participating countries. Events in the list were not specifically based on theoretical underpinnings but reflective of the intentions of all Collaborating Centers from a highly commonsense viewpoint. Some indistinct events with multiple dimensions, overlaps, or similarities with other events that cannot always be categorized into a single event also required a rigorous manipulation. The period was set at 3 months retrospective from the initial date of the study, and it was necessary to set each initial date accurately. This approach applies to the determination of date of each life event experienced within 3 months retrospective from the operationally determined date of onset or relapse.

For each life event identified on the list, *degree of impact* is rated on a 7-point scale from mild or no impact=0, moderate impact=3, and severe or marked impact=7, according to common cultural norms and frameworks and with respect to whether or not the life event is considered to result in *prolonged and substantial changes*. As the last step, *association between schizophrenia* and the life event load was examined. Relevant events were then categorized into *independent life event=0*, *possible independent event=3*, or *probably illness related event=7*.

For consistent operation of the above procedures, the participating Collaborating Centers were repeatedly trained on assessment case practice with the use of multiple descriptive life event reports prepared by the WHO/HQ before and during the study. However, despite these practices, it was more difficult to improve consistency in assessment results with LES than PSE.

Table 5.27 lists the life event models roughly divided into corresponding fields, and, as seen from the item numbers in the right column of each line, the semi-structured interview method is used to assess 70 events as a whole. On the assessment summary form, presence/absence and number of events, date of each event, and degree of impact as well as illness-independent levels are input as codes.

We will now look at the case example of specific assessments.

A male patient, who was 26 years old when he visited a psychiatrist for his first consultation. He was born as the sixth of seven children in the catchment area and had no familial susceptibility to psychiatric disorders. He was an average student in primary and junior high school; his performance was also average at industrial high school. Immediately after graduation from high school, he entered into a company located in another prefecture, but quit the job and returned home after having a hard time adjusting. He then took on a position in a local factory, and for several years he was often out of town on business to neighboring cities. In February 1979, he was on a long business trip in another prefecture, and there he was overwhelmed with such a busy schedule as to be able to see his parents only occasionally. When he was one time able to come home from another business trip destination at the end of March, his parents became aware that he was sighing with continuing insomnia. At the end of April, he was worrying about a trivial failure (trivial for his peers and supervisors) and started to blame himself for having troubled his company.

Around February in the same year, he informed his parents of his plan to marry a woman with whom he had been in a steady relationship for about 2 years, and he introduced her to his parents at the beginning of March. On April 4, engagement

Table 5.27 Types of life events [21]

I. Personal	
A. Move (01)	01. Change of residence?
B. Health (02)	02. Onset of physical illness, disability, or suffering accident?
C. Education (03–05)	03. Begin or change educational programs?
	04. Cease to participate in educational program?
	05. Onset of education related crises?
D. Criminal or legal affair s(06 and 07)	06. Contact with police courts or other legal authorities?
	07. Legal or criminal offenses committed against the subject?
II. Livelihood	*If the subject has been exclusively employed at non wage labor during the relevant period "CUT-OUT" and go on to item no. 16*
A. Paid employment (08–15)	08. Start work for the first time or return to work after substantial break?
	09. Job promotion?
	10. Job demotion?
	11. Change in line of work, hours, or working conditions?
	12. Retirement?
	13. Fired or permanently dismissed?
	14. Substantial business failure?
	15. Unemployed for at least one month?
B. Unpaid employment (16–19)	*If the subject has been exclusively employed at a paid job during the relevant period, "CUT-OUT" and go on to the item no. 20*
	16. Unable to work for one month or more?
	17. Onset of negative economic changes?
	18. Significant economic improvements?
	19. Change in line of work, hours, or working conditions?
C. Public assistance (20)	20. Difficulties with benefits from private or public insurance?
D. Family budget (21 and 22)	21. Onset of other financial difficulties—e.g., housing problems?
	22. Substantial financial improvements?
III. Family/household	*If the subject was not married or living in a common-law union during the relevant period, "CUT-OUT" and go on to the item no. 30*
A. Family after marriage	
1. Marriage (23–29)	23. Death of spouse?
	24. Change in marital status?
	25. Onset of physical illness, disability, or accident to spouse?
	26. Onset of mental illness in spouse?
	27. Spouse either started or stopped working?
	28. Emotional crises with spouse
	29. Pregnancy, miscarriage, stillbirth, abortion, sterilization, or other kinds of sexual difficulties?
2. Children (30–36)	*If the subject has never had any natural, step, adoptive, foster, or surrogate children, "CUT-OUT" and go on to the item no. 37*
	30. Children or grandchildren born?
	31. Children or grandchildren died?
	32. Onset of physical illness, disability, or accident to children?
	33. Onset of mental illness in children?
	34. Onset of behavioral problems with children?
	35. Other exits of children from household?
	36. Other (re)entries of children into household?
B. Family before marriage	37. Death of father and/or mother?
1. Parents (37–40)	*If the subject's parents were both deceased during the relevant period, "CUT-OUT" and go on to the item no. 41*
	38. Onset of physical illness, disability, or accident to father or mother?
	39. Onset of mental illness in father or mother?
	40. Emotional crises with father and/or mother?

(continued)

Table 5.27 (continued)

2. Siblings (41–46)	*If the subject has no natural, step-, adoptive, foster, or surrogate siblings, "CUT-OUT" and go on to the item no. 46*
	41. Death of sibling(s)?
	42. Exit of sibling(s) from household?
	43. Birth or (re)entry of sibling(s) into the household?
	44. Onset of physical illness, disability, or accident to sibling(s)?
	45. Onset of mental illness in sibling(s)?
	46. Emotional crises with sibling(s)?
C. Current family	
1. Birth and death (47 and 48)	47. Other births in household?
	48. Other deaths (including pets) in household?
2. Marital status (49)	49. Change in marital status of household member(s)?
3. Health (50 and 51)	50. Onset of physical illness, disability, or accident to other household member(s)?
	51 Onset of mental illness in other household member(s)?
4. Criminal or legal offenses (52 and 53)	52. Contact with the police, courts, or other legal authorities by household member(s)?
	53. Criminal or legal offenses committed against household member(s)?
5. Entry/exit (54 and 55)	54. Other (re)entries into the household?
	55. Other exits from the household?
6. Emotional crisis (56)	56. Emotional crises with other member(s) of the household?
IV. Social network	*If the subject was married throughout the relevant period, "CUT-OUT" and go on to the item no. 61*
A. Formal engagement/ romantic relationship	
1. New engagement/ romantic relationship (57 and 58)	57. Subject became married?
	58. Subject began romantic relationship?
2. Termination of engagement/ romantic relationship (59 and 60)	59. Subject broke-off engagement?
	60. Subject terminated romantic relationship?
3. Emotional crisis (61)	61. Other crises in a romantic or sexual relationship?
B. General relationships	
1. Termination of an important relationship (62)	62. Termination of an important social relationship?
2. Health (63)	63. Onset of physical illness, disability, accident, or episode of mental illness to important member(s) of social network?
3. Organized social activity (64)	64. Onset of significant changes in organized social activities?
4. Emotional crisis (65)	65. Emotional crises occurring with important member(s) of social network?
V. Additional items	
A. Positive/negative goal achievement (66 and 67)	66. Any valued goal fulfillments?
	67 Any valued goal disappointments?
B. Forecasts of future changes (68)	68. Any forecasts of important future changes?
C. Disturbing news/ revelation (69)	69. Any disturbing revelations about spouse, children, close relatives, intimate friends, or other household member(s)?
D. Other life events (70)	70. Any other significant events?

Partially modified

Table 5.28 A model and summary of life events

Time of onset: May 10, 1979

Life events	Item no.	Date	Impact	Association with the disease	Remarks
1. A long business trip to other prefecture	11	1 0	0	0	Many business trips for several days after the first long business trip for him
2. Mistakes on the job	14?	0 3	3	3	Details were unknown but it seems true that he made a mistake
3. He got married and went on a honeymoon	57	0 2	7	0	A series of events from an engagement ceremony to a honeymoon
4. He made a new home.	01	0 1	3	0	He settled in a new home

Table 5.29 The number of life events 3 months prior to onset of the disease in the patients (%)

No. of life events	Male	Female	All
0 (none)	9.1	19.4	13.8
1–2	43.2	27.8	36.3
3–4	34.1	38.9	36.3
5–	13.6	13.8	13.8

presents were exchanged in a customary ceremony. On April 16 and at the end of April, the place for the wedding and a new house were found, respectively. Immediately after the wedding ceremony on May 6, the patient and his wife went on a honeymoon, and after returning on May 10, they settled into the new house. While on honeymoon, the patient was wreaking his anger on his wife, asserting repeatedly that someone was hearing their conversations, and he was getting significantly less and less sleep. After they were back from the honeymoon, he started talking about his fear that eavesdropping equipment was installed in the clock given to them by the company as a wedding gift and thus to be disposed of immediately and that he was being pursued by a detective agency, and so on. On May 31, he visited a psychiatric practice in the catchment area.

Table 5.28 is a complete descriptive example of the assessment summary. The guide defines explicitly how to describe the assessment results of each event; however, I acutely felt the need for further deliberate efforts toward a completely consistent and conclusive assessment result.

Table 5.29 summarizes life events occurring before disease onset based on this case history. The patient had experienced four life events by the time of onset.

We had the same pattern of LES interviews with 80 of 107 subjects. Table 5.30 indicates the number of life events of all the interviewees. Of all subjects, 11 (13.8 %)

Table 5.30 Total number of life events by sex and per patient in association with the disease

	Male		Female		All	
	Total (%)	Per patient	Total (%)	Per patient	Total (%)	Per patient
Independent life event	89 (80.9)	2.02	73 (78.5)	2.03	162 (79.8)	2.03
Probably independent life event	16 (14.5)	0.36	14 (15.8)	0.39	30 (14.8)	0.38
Probably illness-related life event	5 (4.5)	0.11	6 (6.5)	0.17	11 (5.4)	0.14
Total	110 (100)	2.50	93 (100)	2.58	203 (100)	2.54

Table 5.31 Impact of independent life events

Impact	Male	Female	All
Mild or no impact	15 (16.9)	3 (4.1)	18 (11.1)
Moderate impact	52 (58.4)	48 (65.8)	100 (61.7)
Severe or marked impact	22 (24.7)	22 (30.1)	44 (27.2)
Total	89	73	162

had no event within 3 months before onset. More cases with no event before onset were clearly observed in women than in men. On the other hand, most subjects overall with *one or more events*, no matter the gender, experienced more than one event, with an average number of events of 2.54, as Table 5.30 shows.

Looking at life events experienced within 3 months before onset from the viewpoint of illness/schizophrenia-independent levels, as Table 5.30 illustrates, the majority of events were categorized as *independent life events*, while about 80 % of all events experienced were independent life events. The average number of events per individual was 2.03, which was almost equally distributed in men and women. The other two categories—*possible independent event* and *probably illness-related event*—were not addressed in this life event study, because they are suspected of having association with schizophrenia.

Regarding degree of impact, as Table 5.31 shows, *moderate impact* and *severe and marked impact* comprised the largest portion, 89.9 %.

The specific content of schizophrenia-inducing events as suggested in Table 5.32, work- or education-related events and marriage and romantic relationship-related events, were clearly predominant in men and women, respectively.

We then took a look at the role each life event plays on a mode of onset basis. The same definition as in Fig. 5.4 was used for mode of onset by categorizing life events into acute, subacute, and chronic onset and dividing the 3-month period before onset into three stages of 4 weeks. Then the number of patients who experienced more than one event within each stage was compared.

Table 5.32 Life events per gender (%)

Events	Male (n=44)	Female (n=36)	All (n=80)
Work-/education	43* (53.8)	25* (40.3)	68 (47.9)
Health/illness	6 (7.5)	2 (3.2)	8 (5.6)
Marriage/romantic relationship	5** (6.3)	16** (25.8)	21 (15.5)
Economic/family budget	2 (2.5)	3 (4.8)	5 (4.2)
Move/long business trip	5 (6.3)	5 (8.1)	10 (7.0)
Human relations	14 (17.5)	7 (11.3)	21 (13.4)
Others	5 (6.3)	4 (6.5)	9 (6.3)
Total	80 (100)	62 (100)	142 (100)

*$p<0.1$; **$p<0.005$

Table 5.33 Mode of onset in patients who experienced at least one life event before the onset (%)

	9–12 weeks before	5–8 weeks before	1–4 weeks before
Acute/subacute (n=50)	42.0	52.0	68.0
Chronic (n=25)	16.0	24.0	36.0
All (n=75)	33.3	47.7	57.3

Of 107 subjects, 75 were available for assessment of onset pattern, and among these, 27, 23, and 25 cases were categorized into acute, subacute, and chronic onset, respectively. Bringing the acute and subacute onset patterns into one category, mode of onset was ultimately divided broadly into 2 categories of acute/subacute with 50 subjects and chronic with 25 subjects. As Table 5.33 suggests, acute/subacute schizophrenia patients definitely experienced more life events than chronic patients in stage I (1–4 weeks before onset), as well as in stages II and III (5–8 weeks and 9–12 weeks before onset). This may suggest that the onset of acute/subacute schizophrenia is more intensively related to life events.

The same analysis was performed by subtypes of schizophrenia, such as paranoid and non-paranoid. Since the results showed no particular significance, detailed analysis data are omitted here, concluding that data by onset pattern have more implication for life events than those by subtype.

An association between life events such as loss experience owing to separation or bereavement from family during childhood and various loss experiences that follow in later life has long been known to increase liability to depression, and it is also known that hypotheses such as *Onset models in Life Events and Illness* constructed by Brown and Harris are available [22]. But the specificity of life events related to schizophrenia onset remains poorly understood as a constituent of the vulnerability-stress-coping model, as earlier mentioned.

However, as far as the association between life event load and subsequent lifetime course has been demonstrated, the findings obtained will be presented in the following section, *Natural history and outcome*.

5.5.4 Natural History and Outcome (in the Short and Long Term), and Associated Factors

Natural history studies and outcome studies have a pivotal aim in clinical studies. From the standpoint of a provider of mental health services, such studies should not only be taken into consideration as one of the important aspects through the clinical observational period of diseases/syndromes, but they are also indispensable to verifying the effectiveness of the treatment given. In addition, it surely serves as a vital means to ask the patient and his or her family for their proactive participation or cooperation in treatments and care. Providing them with appropriate information should be a prerequisite to the above purpose, but in a routine medical practice, proper and sufficient information is not always available for every treatment. The physician and medical scientist as provider of health-care services should accordingly keep in mind that what is asked of them is to constantly establish higher-level evidence.

Several findings and reports on course and outcome of schizophrenia from the Burghölzli study conducted by Bleuler in 1972 [23], the Lausanne study conducted by Chiompi in 1976 [24], the Bonn study conducted by Huber in 1975 [25] and the like are widely known. Meanwhile, in Japan, about 20 study reports from Hayashi and Akimoto [26] and other study groups, including the Gunma University group which conducted a long-term follow-up study, have been accumulated. Through such reports, *the one-third hypothesis*, as well as the *scissors-form phenomenon*, which was established by the Gunma University group and explains patients' outcome patterns polarized as either living a self-supporting life again or remaining in prolonged hospitalization, have been generalized.

Giving further insights into the natural history of the disease using the above terms, a recent focus has been placed on events such as *remission* and *recovery*, and definitions for them are also being proposed. The presence or absence of resolution will be assessed immediately after treatment starts, followed by the assessment of remission possibility, before the final assessment of recovery. Simulation of such a process is clinically quite meaningful.

As one criterion for remission, the Positive and Negative Symptoms Scale (PANSS), established by Andreasen et al. [27], is widely known as a medical scale to assess remission based on eight PANSS items. The scale consists of P1 (delusions), G9 (unusual thought content), and P3 (hallucinatory behavior) as general psychopathology symptom items; P2 (conceptual disorganization) and G5 (mannerisms/posturing) as positive symptom items; and N1 (blunted affect), N4 (social withdrawal) and N6 (lack of spontaneity) as negative symptom items. Remission is defined as a score of 3 (mild) or less (minimum/absent) for each item (each item is scored on the same seven-point severity scale), maintained over 6-month period.

Recovery can be defined as a state of remission maintained for a longer period, in which almost all symptoms have diminished. While more and more foreign study results using these criteria have been reported, in Japan only review papers are available. In any case, it may be high time that such particular terms and concepts should be deliberately examined to be established as EBM in Japan.

A sample of 107 first-episode schizophrenia patients had been continuously monitored in the DOSMeD study at 1-, 2-, 5-, 10-, and 15-year follow-up periods, as well as at a symptomatic relapse after the initial consultation. As the final follow-up, an ultra-long-term outcome study is currently under way to monitor the outcomes at a 20-year period. As shown in Table 5.34, diligent follow-up results have been consecutively reported by the Gunma University group. DOSMeD data to be obtained is also expected to be consistent with their achievements. It is important to note that the criteria used in our follow-up studies differ slightly from those used in aforementioned foreign studies and studies conducted by the Gunma University group.

Table 5.35 summarizes the follow-up data from 2- to 15-year periods. The number of subjects included in the DOSMeD study was initially 107, but while we garnered cooperation from 73 of them for the follow-up at a 2-year period, only 65–68 (included death) remained available for further follow-ups, which unfortunately made a definitely smaller cohort than that of the Gunma University group. This can be attributable to inclusion criteria for the DOSMeD study being satisfied by those who had the first episode of schizophrenia after moving into the catchment area in Nagasaki City. In other words, schizophrenia patients who had been treated in 30 psychiatric hospitals in Nagasaki City and surrounding areas have segued into the follow-up phase without being transferred to any designated university hospitals or other institutions. From another viewpoint, the outcome data should, however, reflect a general treatment pattern given to schizophrenia patients in Nagasaki.

First, the 107 subjects were monitored for frequency of relapse at 1- and 2-year follow-up periods at the Collaborating Center in Nagasaki. For these follow-ups, relapse was limited to first onset of psychotic state after complete or partial remission, and 38 % had one or two relapses by the 2-year follow up period, increasing to 57 % and 70 % at 5- and 10-year periods, respectively. This was a fresh reminder of a higher relapse rate of first episode of schizophrenia than our initial estimation.

The association between life events before onset and treatment withdrawal has been controversially considered to be a risk factor related to relapse. This reflects the assessment results repeated on the association between life events and onset of disorders based on the earlier-mentioned rigorous criteria. Each assessment focused on content, frequency, and degree of impact of life events experienced within the operationally determined 3-month period before onset. The same endpoints were adopted for assessment of relapse-associated life events. Also in the follow-up interviews at 1- and 2-year periods, life events that occurred within 3 months before each period were examined. As Fig. 5.5 illustrates, an obvious clear difference was to be found in per-patient number of *independent life events*, which was shown to be 2.03 on average within 3 months before onset, but reduced to 1.95 and 1.10 within 3 months before relapse and regular follow-ups, respectively. The number of life events within 3 months decreased in the order of before onset, relapse, and regular follow-ups, and, in particular, the relationship between distinct onset of symptoms and increased life events has been demonstrated to be a contributing factor.

Figure 5.6 shows the findings on the relationship between treatment withdrawal and life events, indicating relapse rates of schizophrenia within 2 years after onset.

Table 5.34 Results of the long-term follow-up by the Gunma University group

Duration of the follow-up		2–7 years	5–10 years	15–20 years	21–27 years	Mean 35 years
Start date of the follow-up		1964.11	1967.10	1978.01	1984.01	1996.01
Number of cases		$n=124$	$n=120$	$n=112$	$n=105$	$n=127$
Social adjustment	Self-supporting	45.2 %	55.7 %	44.0 %	46.7 %	42.5 %
	Semi-self-supporting	13.9	10.6	11.8	7.6	8.0
	Adjustable in the family	12.2	7.1	7.6	11.4	12.6
	Social maladjustment	5.2	3.5	3.3	2.9	1.1
	Hospitalization	23.5	23.0	33.4	31.4	34.5
	Death (n)	(7)	(7)	(19)	(25)	(41)
Course of social adjustment	Continuously good	18.5	16.8	31.1	Recovered	
	Gradually good	12.9	17.7	25.6	31.4	32.5
	Generally good	37.9	38.9	5.6	Improved	
	Generally poor	17.7	12.4	2.2	45.7	43.4
	Gradually poor	6.5	8.9	2.2	Not cured	
	Continuously poor	6.5	5.3	33.3	22.9	24.1

At the initial assessment in 1958, $n=140$, assessed with the Eguma assessment scale, 1963

Table 5.35 Summaries of the DOSMeD follow-up

Duration of the follow-up		2-year	5-year	10-year	15-year
Period		1981–1982	1984–1985	1989–1990	1994–1995
Terminated cases (incl. death)		73	66	66	65
Accumulated deaths (% of all cases in the follow-up)		1 (1.4 %)	1 (1.5 %)	5 (7.6 %)	7(10.8 %)
Terminated cases (excl. death)		72	65	61	58
Frequencies of outcomes observed in cases during the follow-up					
Complete remission	w/relapse	6 (8.3 %)	4 (6.2 %)	2(3.3 %)	4 (6.9 %)
	w/o relapse	14 (19.4 %)	15 (23.1 %)	16 (26.2 %)	12 (20.7 %)
Partial remission	w/relapse	13 (18.1 %)	22 (33.8 %)	27 (44.3 %)	28 (48.3 %)
	w/o relapse	15 (20.8 %)	14 (21.5 %)	2 (3.3 %)	0
Maintained psychotic symptoms		24 (33.3 %)	12 (18.5 %)	13 (21.3 %)	12 (20.7 %)
Other				1 (1.6 %)	2 (3.4 %)

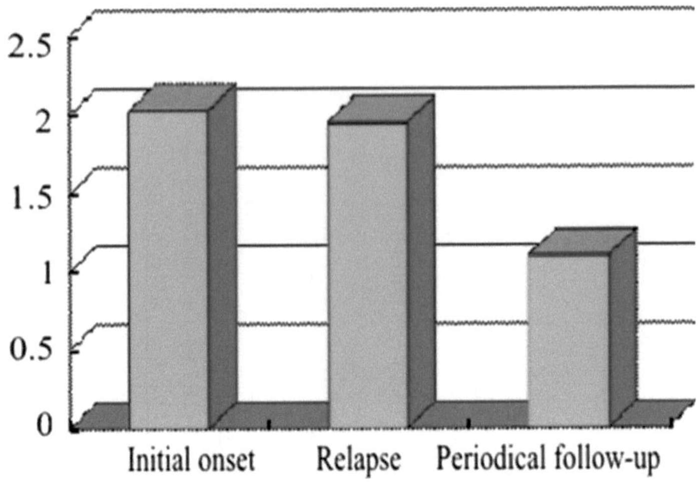

Fig. 5.5 Number of life events at the time of each evaluation (per patient)

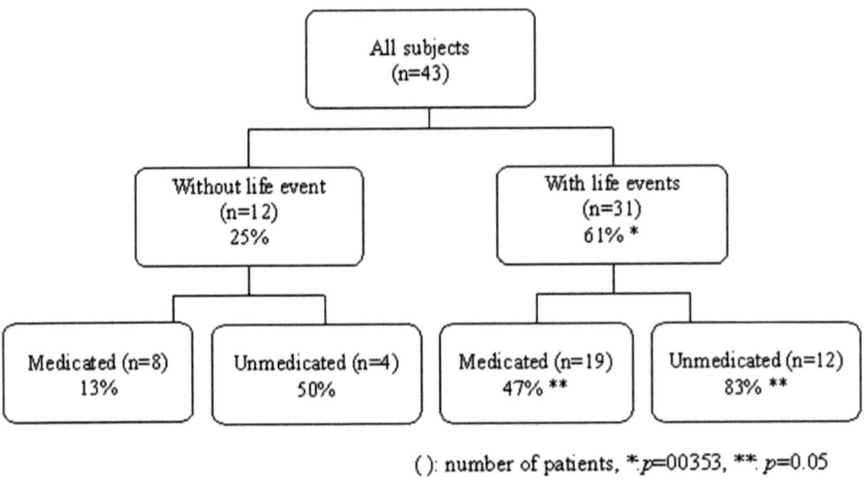

Fig. 5.6 Relationship between life events and medication at the 2-year follow-up period (relapse rate)

Because endpoints of treatment withdrawal and presence/absence of life events were evaluated concurrently, subjects were too diversely categorized. The relapse rate was the highest in the unmedicated patients with life events and the lowest in medicated patients without life events. Even though which endpoint is more intensively involved in relapse is still to be ascertained, this result implies that continuous medication, as commonly known, can contribute to a reduced relapse rate.

Generally observed treatment courses and subject outcomes can be summarized as follows.

The final follow-up period of the DOSMeD study was initially set to 5 years. But upon request from the participating foreign Collaborating Centers, the International Study of Schizophrenia (ISoS) was setup to examine the course and outcome of schizophrenia by comparing the follow-up data of DOSMeD study subjects to those of subjects included in foreign follow-up studies that used standardized assessment instruments.

I would now like to explain about the terms defined in our study to identify disease courses. The psychotic state is a state in which a patient keeps satisfying two or more criteria shown in Table 5.5 until the next follow-up period; partial remission is a state in which a patient is present with nonpsychotic symptoms, such as depression and neurotic symptoms or negative symptoms, that do not satisfy the above inclusion criterion (2) over one month or longer; and complete remission is a state in which all the symptoms observed at onset have diminished. As the Table 5.35 shows, disease courses can accordingly be classified into the following five categories: *complete remission after one psychotic state*; *partial remission after one psychotic state*; *complete remission after several psychotic states*; *partial remission after several psychiatric stages*; and *maintained psychotic state*. Social-adjustability-related outcomes were evaluated using the WHO Psychiatric Disability Assessment Schedule (WHO/DAS), developed in 1988 in accordance with the draft of the International Classification of Impairments, Disabilities and Handicaps (ICIDH), which was promulgated by WHO in 1980. Meanwhile, DAS has been elaborated into the current International Classification of Functioning, Disability and Health, ICF, 2001 (the Japanese version was issued in 2002) after two revisions of WHO/DAS-II (translated into Japanese in 1998) and ICIDH-2 (1999; the Japanese version was published in 2000). DAS usually assesses disease courses on a 7-grade scale, but the DOSMeD study used scales of *very satisfactory (the same level as a healthy individual)*, *satisfactory*, *poor*, *very poor*, and *under hospitalization*.

As far as life prognosis is concerned, 7 deaths were confirmed at the 15-year follow-up period after onset, which makes 10.8 % of the subjects that were available for regular follow-ups, 6.5 % of all the included subjects. More than 10 years later, 12 deaths were confirmed at the 28-year period after onset, which makes 26.1 % in 46 subjects who were available for regular follow-ups, 11.2 % of all the included subjects. Among those, six deaths were identified as apparent suicides (5.6 % of all the included subjects, three cases each in men and women), three deaths were due to drowning, one was an apparent suicide due to death-leap and elaborated self-injury behaviors, and another death was listed as due to unspecified reason. Deaths other than suicide were induced by physical diseases such as cancer (two cases) and respiratory tract infection (two cases).

As one of the long-term observational studies on schizophrenia, the Iowa study, conducted by Tuang and Winokur et al. in the USA [28], is well known for its prominent report of frequency of death. In this report, death rate from suicide or in a freak accident was shown to be approximately10%, which is almost the same high rate as seen in mood disorder patients compared to the general population. In Japan,

Ichimiya et al. reported 8 deaths in 129 subjects (6.2 %) in their 20- and 44-year long-term follow-up data conducted by the Gunma University group [29, 30]. In mood disorder patients, death from suicide is usually seen at a relatively early stage, within 10 years after onset, but some have reported that a characteristic of schizophrenia is that death from suicide is evenly distributed throughout the given follow-up period. A continued up-close look at follow-up results may accumulate more data on death from suicide.

As far as outcome is concerned, symptomatic remission of the patient should be assessed, as mentioned earlier. Outcome is also assessed according to, for example, patients' comprehensive social adjustability and social involvement within 1–2 years before each assessment.

In the DOSMeD study, complete and partial remission as well as maintained psychotic symptoms were assessed not based on the aforementioned medical scale established by Andreasen et al. [31], but with our own criteria from the viewpoint of long-term follow-up of symptomatic remission. As already seen in Table 5.35, 27–30 % showed complete remission (in particular, complete remission after one episode remains at 4–8 %); 39–55 % partial remission in psychotic symptoms were being maintained in 19–33 % at the first to fourth follow-up periods. No remarkable difference was observed in frequency at each follow-up period, but the partial remission rate at the 5-year period appeared to have increased by the corresponding rate decrease in maintained symptoms in the same period. But the likelihood that this difference observed at the 5-year period resulted from the use of inconsistent assessment instruments cannot be excluded. Fewer than 40 subjects only experienced one symptomatic phase from the first onset through 5 years, and fewer than 30 % of them showed a complete remission, but at the follow-up periods thereafter, only 7–8 experienced one episode and eventually relapsed.

When categorizing symptomatic outcomes seen within 2 years before the 15-year follow-up period into *absence of symptoms*, *presence of intermittent symptoms*, *presence of persistent symptoms*, and *not classifiable*, the distribution was found to be 24 %, 16 %, 53 %, and 6 %, respectively, suggesting that about half of schizophrenia patients in Nagasaki are still afflicted with chronic psychotic symptoms 15 years after onset.

Table 5.36 shows patients' comprehensive social adjustability at the 15-year period after onset on a 6-grade scale of *very satisfactory*, *satisfactory*, *relatively*

Table 5.36 Comprehensive social adjustability in the 15-year follow-up (WHO/DAS-S) [31]

Conditions	Cases (%)	Good/poor	Favorable/unfavorable
Very satisfactory	6 (10.3)	Good	Favorable:
Satisfactory	8 (13.8)	14 (24.1 %)	23 (39.7 %)
Relatively satisfactory	9 (15.5)		
Unsatisfactory	19 (32.8)		Unfavorable:
Poor	11 (19.0)	Poor	32 (55.2 %)
Seriously poor	2 (3.4)	13 (22.4)	
Not assessable	3 (5.2)		
Total	58		

Table 5.37 Long-term outcomes in the patients with initial schizophrenic episode

Study reports	Mean follow-up period (year)	Cases	Clinically favorable outcome (%)	Clinically unfavorable outcome (%)	Favorable social adjustment (%)
Ciompi (1980)	37	289	27	42	39
Bleuler (1978)	23	208	20	24	51
Bland/Orne (1978)	14	90	26	37	65
Salokanngas (1983)	8	161	26	24	69
Shepherd (1989)	5	49	22	35	45
Nagasaki (1998)	15	65	28	21	40

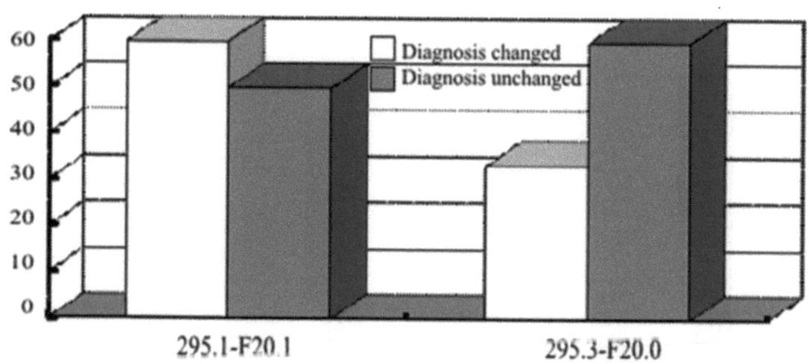

Fig. 5.7 Assessment of (poor) outcomes at the 15-year follow-up period after re-categorization from ICD-9 to ICD-10 (%)

satisfactory, *unsatisfactory*, *poor*, and *extremely* or *seriously poor*. If these are roughly converged into two categories of *satisfactory* and *poor*, they account for 40 % and 55 %, respectively (in 5 %, assessment was not possible), and, among all, 24 % showed an apparently satisfactory adjustability and almost the same rate of 22 % apparently poor adjustability. Table 5.37 summarizes the accumulated study reports on outcomes of first episode of schizophrenia with reference to symptoms and social adjustability. As far as this table suggests, it may be concluded that there was no notable change in outcomes of patients for the past 20 years.

However, considering the fact that the clinical status improved similarly, social adjustability data is considered reflective of the difference in criteria used in the assessments.

Here we will briefly touch on potential factors that affect such outcomes. The following are our study results and may thus require additional studies for generalization.

As mentioned earlier, ICD was revised in the course of the DOSMeD study, and, in the revised ICD, standardized diagnostic criteria were established and the included subjects re-categorized into relevant schizophrenia subtypes accordingly. Figure 5.7 indicates the frequency of poor outcome assessed in the hebephrenic type (ICD-10/F-20.1) and the paranoid type (ICD-10/F-20.0) at the 15-year follow-up period. Poor outcomes were more frequently observed in subjects diagnosed with hebephrenic

Fig. 5.8 Age groups of onset and complete remission rate (%) at each follow-up period

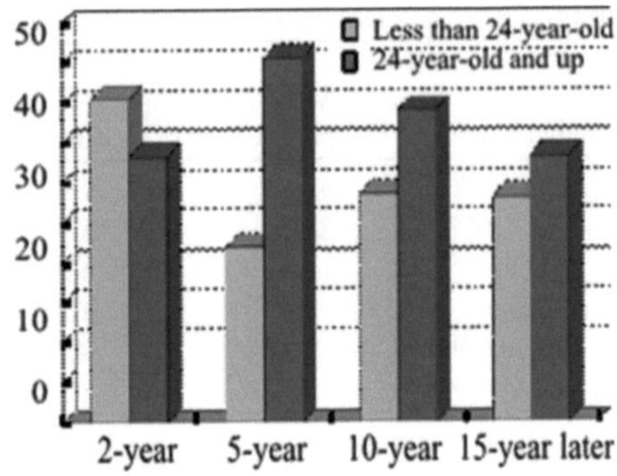

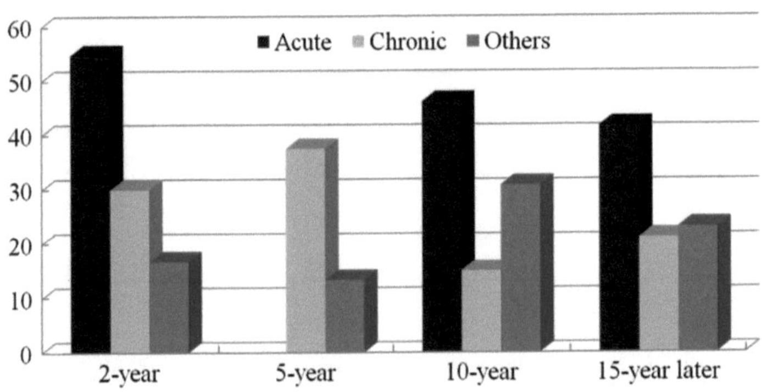

Fig. 5.9 Onset patterns and complete remission rate (%) at each follow-up period

schizophrenia according to both the ICD-9 and ICD-10, that is, the diagnosed subtype remained unchanged after the use of revised diagnostic criteria, whereas subjects newly categorized into paranoid schizophrenia according to ICD-10 showed poor outcomes more frequently. This implies that the effect of the change in categorization on outcome varies according to disease subtypes.

The factors proven to be associated with outcome include age of onset, mode of onset, social adjustability before onset especially in childhood and early adolescence, presence or absence of life events at onset and relapse, and delay in treatment start. We will see the effect of such factors on long-term outcomes in the DOSMeD study results.

Figure 5.8 illustrates the effect of age of onset on outcome after 2, 5, 10, and 15 years, when comparing two subject groups with age of onset of 24 or less and over 25 years. In the short-term period, for example, after 2 years of follow-up, the group with the younger age of onset showed a higher complete remission rate, but the difference between the two groups became smaller in long-term outcomes.

Figure 5.9 illustrates the effect of acute or chronic onset pattern on outcome. As previously mentioned, this comparison is based on onset patterns determined

Table 5.38 Adjustment in childhood and early adolescence and the outcomes after 10 years

	Some adjustability problem in childhood (%)	Some adjustability problem in adolescence (%)	Some problem in either period (%)
Favorable outcome group	53.8	46.2	69.2
Unfavorable outcome group	73.1	65.4	84.6
Total	66.7	59.0	84.6

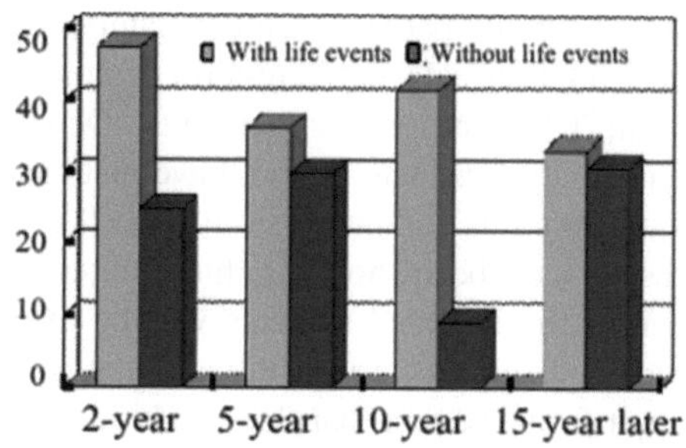

Fig. 5.10 Presence of *life events within 3 months before onset* and complete remission rate

operationally. Onset patterns seemed to affect outcome less and less over time, but chances of acute onset for overall better outcome have been shown to persist throughout the study period.

For assessment of social adjustment, subjects were interviewed on 22 and 20 items with respect to adjustability in childhood and early adolescence, respectively, based on a predefined instrument at the initial assessment. Results from the 5- and 10-year follow-ups showed no significant difference in potential predictors, but a closer look at outcome at the 10-year period in Table 5.38 suggests that it does not make a statistically significant difference, though subjects who had problems in their childhood or early adolescence with adjustability, or with any other problems in either of the above time periods, may have a poor outcome, with a poor outcome ratio of 66.7 %, 59.0 %, and 84.6 %, respectively.

Lastly, as shown in Fig. 5.8 earlier in conjunction with life events, Fig. 5.10 suggests the successive effect of life events on long-term outcomes. A higher complete remission rate was observed in subjects who had experienced more than several life events within 3 months before onset, but at the 15-year period there seemed to be less causality between number of the patient with life events and complete remission. Just to see such causalities, we have to admit that a large variability in data is seen due to the small sample used, so further validation with a larger sample may be required.

An ultra-long-term outcome study with a study period of more than 20 years from the onset of initial episodes is currently underway in these 107 subjects. The current address of many subjects remains unknown due to the complex procedure required to get in touch with them as well as to obtain their consent to participate voluntarily in this study, even if we succeed in identifying their place of residence.

Therefore, this study cannot be said to be proceeding smoothly. It should take some more time for significant findings.

Despite the common recognition of the needs for such long-term follow-up studies, it is not easy to continue them. I presume that the time to withdraw from our study is also nearing, but I hope it will happen as late as possible. Recently, it is becoming a common tendency to want a quick result, and this creates a tendency to defer from continuing such long-lasting studies. But it is necessary to track the course of not only some cases shown to be interesting but also a large enough number of control subjects over quite a long term, to speak definitively on the lifetime outcome of schizophrenia patients. Unfortunately, fewer and fewer studies can in fact afford such a large-scale and long-term design.

In 2010, junior researchers of the Department of Neuropsychiatry at the Nagasaki University Graduate School have resumed the 28-year follow-up in the DOSMeD study. As this follow-up study is still ongoing, only a small fraction of collected results have been shown in this chapter. After 28 years from the onset, the average age of the included subjects will be over 50. The fact that 12 deaths in the 107 included subjects are identified makes us aware of the significant passage of time, apart from the fact that this morbidity is higher than the death rate in the general population.

References

1. Tatetsu S. Epidemiology. In: The group publishing the book to commemorate Prof. Y. Uchimura, editor. Personality and achievement. Tokyo: Sozo Publishing; 1982 (in Japanese).
2. Takahashi R, Harada K, Nakane Y, Matsunaga F, Kawasaki N. Physical and mental developments in childhood and psychiatric diseases. Clin Psychiatry. 1976;18:853–63 (in Japanese).
3. Nakane Y, Fujii I, Ohta Y, Morita T, Takahashi R. Physical and mental development in childhood and risk of schizophrenia in later life. Folia Psychiatr Neurol Jpn. 1978;32:63–75.
4. Imamura Y, Nakane Y, Ohta Y, Kondo H. Lifetime prevalence of schizophrenia among individuals prenatally exposed to atomic bomb radiation in Nagasaki City. Acta Psychiatr Scand. 1999;100:344–9.
5. Nakane Y. Epidemiology in psychiatry. In: Kakeda K, Shimazono Y, Okuma T, Takahashi R, Hozaki H, editors. Handbook of modern psychiatry' supplement for '87-A. Tokyo: Nakayama Shoten; 1987. p. 31–57 (in Japanese).
6. Nakane Y, Takahashi R, Tominaga Y, et al. The incidence rate study of schizophrenia in Nagasaki city. Clin Psychiatry. 1985;27:771–81 (in Japanese).
7. Nakane Y, Takahashi R, Ohta Y. Morbid risk of schizophrenia in Japan. Clin Psychiatry. 1986;28:421–6 (in Japanese).
8. Saha S, Chant DC, Welham JL, McGrath JJ. The incidence and prevalence of schizophrenia varies with latitude. Acta Psychiatr Scand. 2006;114:36–9.
9. Sugasaki H, Michitsuji S, Ohta Y, Nakane Y. International collaborative studies of schizophrenia: International Pilot Study of Schizophrenia (IPSS) and WHO Collaborative Study on the Determinants of Outcome of Severe Mental Disorder (DOSMeD). Psychiatr Diagn. 1994;5:39–53 (in Japanese).
10. Bøjholm S, Strömgren E. Prevalence of schizophrenia on the island of Bornholm in 1935 and in 1983. In: Sartorius N, Nielsen JA, Strömgren E, editors. Changes in frequency of mental disorder over time. Acta Psychiatr Scand. 1989;9(Suppl. 348):157–66.

11. Helgason T, Magnussen H. The first 80 years of life. A psychiatric epidemiological study. In: Sartorius N, Nielsen JA, Strömgren E, editors. Changes in frequency of mental disorder over time. Acta Psychiatr Scand. 1989;79(Suppl. 348):85–94.
12. Hagnell O. Repeated incidence and prevalence studies of mental disorders in a total population followed during 25 years. The Lundby Study, Sweden. In: Sartorius N, Nielsen JA, Strömgren E, editors. Changes in frequency of mental disorder over time. Acta Psychiatr Scand. 1989;79(Suppl. 348):61–77.
13. Jablensky A, Sartorius N, Ernberg G, et al. Schizophrenia: manifestations, incidence and course in different cultures. A World Health Organization Ten-Country Study. Psychological Medicine Monograph 20. Cambridge: Cambridge University Press; 1992. p. 1–97.
14. Faris REL, Dunham HW. Mental disorders in urban areas, an ecological study of schizophrenia and other psychoses. New York: Häfner Publishing Co. Inc.; 1939 (reprinted in 1960 by University of Chicago Press, Chicago).
15. Takemoto T, Nakane Y, Nishihara J, Ohta Y, Moji K, Izumi T. Human ecological system in the city and its health consequences. Jpn J Soc Psychiatry. 1988;11:259–67 (in Japanese).
16. Nakane Y, Takada K, Yoshitake K, Hatada K. Chapter 15, DOSMeD: Nagasaki, Japan. In: Hopper K, Harrison G, Janca A, Sartorius N, editors. Recovery from schizophrenia. London: Oxford University Press; 2007.
17. Nakane Y, Araki K, Ohta Y. So-called "Hören von Stimme in der Form von Rede und Gegenrede" of schizophrenia – from the stand point of transcultural psychiatry. Jpn J Soc Psychiatry. 1983;6:93–9 (in Japanese).
18. Harima H. Clinical psychopathology, new edition (translated from Klinische Psychopathologie by Schneider K, 15 Auflage). Tokyo: Bunko-do; 2007.
19. Nakane Y, Takahashi R. International diagnostic criteria for schizophrenia. Jpn J Clin Psychiatry. 1982;11:1327–35 (in Japanese).
20. Ishizawa M. Study on life events and schizophrenia I. Relationship of life events to onset of illness. Nagasaki Med J. 1983;58:264–88 (in Japanese).
21. Department of Neuro-psychiatry. Japanese version of assessment guideline for WHO/life event schedule. Nagasaki: Nagasaki University School of Medicine; 1983 (private printing, in Japanese).
22. Brown GW, Harris T. Social origins of depression: a study of psychiatric disorder in women. Cambridge: Tavistock Press; 1978.
23. Bleuler E. Dementia praecox oder Gruppe der Schizophrenien. Leipzig: Franz Deuticke; 1911 (translated into Japanese by Iida M, Shimosaka K, Hozaki H, Yasunaga H, 1974. Igaku-Shoin, Tokyo).
24. Chiompi L. Lebensweg und Alter der Schizophrenen. Eine katamnestische Langzeitstudie bis ins Senium. Berlin: Springer; 1976.
25. Huber G, Gross G, Schüttler R. A long-term follow-up schizophrenia: psychiatric course of illness and prognosis. Acta Psychiatr Scand. 1975;52:49–57.
26. Hayashi A, Akimoto H. Prognosis and treatment of schizophrenia. Psychiatr Neurol Jpn. 1939;43:705–42 (in Japanese).
27. Andreasen NC, Falum M, Arndt S, et al. Positive and negative symptoms: assessment and validity. In: Marneros A, Andreasen NC, Tsuang MT, editors. Negative versus positive schizophrenia. Berlin: Springer; 1991. p. 28–51.
28. Tuang MT, Woolson RF, Flemming JA, et al. Long-term outcome of major psychoses. I. Schizophrenia and affective disorders compared with psychiatrically symptom-free surgical conditions. Arch Gen Psychiatry. 1979;39:1295–301.
29. Ichimiya Y, Ishikawa I, Sakurai N, et al. Outcome of schizophrenia. Psychiatr Neurol Jpn. 1986;88:206–34 (in Japanese).
30. Ogawa K, Miya M, Watari A, et al. A longterm follow-up study of schizophrenia in Japan: with special reference to the course of social adjustment. Br J Psychiatry. 1987;151:758–65.
31. Andreasen NC, Carpenter Jr WT, Kane JM, et al. Remission in schizophrenia; proposed criteria and rationale for consensus. Am J Psychiatry. 2005;162:441–9.

Chapter 6
Studies on Affective Disorders

Over the past decade, more than 30,000 completed suicides have been reported annually. The most frequent factor for suicide has been shown to be a high rate of health problems, more so than personal reasons and financial matters. Since many suicidal individuals were plagued with *depression* above all, treating depression has increasingly been attracting attention as the key to a strategy for suicide prevention. On the other hand, it is also said that depression has proliferated in the general public and is present in atypical forms.

Because our laboratory, led at that time by Professor Ryo Takahashi, had joined the cross-national research study on depression since well before its designation as a World Health Organization (WHO) Collaborating Center for Reference and Research, our laboratory has been continuously involved in symptomatologic and psychosocial research on mood disorders, as shown in Table 5.6 earlier.

6.1 WHO World Mental Health Survey in Japan

Since 2002, the WHO World Mental Health Survey (WHO/WMH-J) has been underway in our laboratory as a state-of-the-art depression study. This international collaborative study is a WHO-driven cross-national epidemiological project implemented in 17 countries around the world. Centers other than Nagasaki in Japan had also contributed to case-finding for this multicenter collaborative study until 2006.

For this study, subjects were randomly selected from among general eligible voters aged 20 or over (18 or over in most foreign countries) living in the catchment areas, who consented to be included in this study. A door-to-door interview was conducted with every subject by nonpsychiatric experts who had been trained in the interview methods used and other procedures for 5 days in advance. As the interview method, WMH adapted the computer-assisted personal interview (CAPI) of the third edition of the Composite International Diagnostic Interview (CIDI) developed by WHO [1]. This CAPI consisted of the following two parts: Part I (interview with all subjects to assess prevalence rate of the intended mental disorders, and their

Y. Nakane, *Progress in Social Psychiatry in Japan: An Approach to Psychiatric Epidemiology*, DOI 10.1007/978-4-431-54103-5_6, © Springer Japan 2012

severity and treatment patterns) and Part II (additional interview with randomly selected subjects to assess mental-health-related factors and perceptions). Through the use of this interview method, not only diagnosis of subjects' mental status, severity, and treatment patterns, but also the level of mental health literacy each subject had was able to be comprehensively understood.

Psychiatric disorders for this research were determined based on the objective of revealing the most commonly referred mental disorders within the scope of mental health among the general public, as well as on epidemiological data pertinent to mental disorder categories that contemporary physicians most frequently encounter in practice. Thus, they did not cover all the mental disorders WHO–CIDI contains, but were limited to so-called common mental diseases. Corresponding disorders in ICD-10F include six *mood disorders*, consisting of F32.2-3 severe depressive episode, F32.1 moderate depressive episode, F32.0 mild depressive episode, F30.1-2 manic episode, F30.0 hypomania, and F34.1 dysthymia; six *neurotic, stress-related, and somatoform disorders*, consisting of F41.0 panic disorders, F40.00 agoraphobia without panic disorder, F40.1 social phobias, F40.2 specific (isolated) phobias, F41.1 generalized anxiety disorder, and F43.1 posttraumatic stress disorder; and four *mental and behavioral disorders due to psychoactive substance use*, consisting of F10.1 harmful use-alcohol, F10.2 alcohol dependence syndrome, F1x.1 harmful use-drugs, and F1x.2 drug dependence syndrome.

Psychotic disorders, such as schizophrenia, and *organic, including symptomatic, mental disorders*, such as dementia, were excluded.

The Nagasaki Collaborating Center spearheaded this field research initiative, which was followed by other regions in Japan. As described later in this chapter, I had anticipated sufficient cooperation from the general public based on my experience with the field studies repeated in atomic bomb survivors, but unfortunately the reality proved different.

Table 6.1 represents the subjects recruited for WHO/WMH-J in six prefectures nationwide [2]. I feel very embarrassed with the very low response rate in Nagasaki in comparison with the rates of other centers. In the wake of this, a strategy was deliberately devised to enhance cooperation in the public and, as a result, 4,134 of the more than 8,000 people who were asked to participate in the study across the nation consented to be included in the survey (average response rate: 55.1 %). These subjects were then interviewed according to the structured interview schedule. This is an unprecedented large-scale field study on mental health in Japan using a personal interview in the general public.

Data on frequency of the earlier-mentioned common mental diseases are already available. However, only the data on mood disorders will be shown here.

As Table 6.2 suggests, lifetime prevalence of all the mood disorders including depression accounts for 7.5 % of the 4,134 subjects surveyed [2]. In other words, nationwide about one in thirteen people are likely to develop mood disorders including depression at least at some point in their life. As a gender-specific pattern, women have about a two times higher prevalence rate than men (9.9 vs. 4.6 lifetime prevalence), as has long been known. By age groups, people 65 or over show the lowest prevalence rate, and the younger the age group is, the higher the rate becomes.

Table 6.1 Subjects of WHO/WMH-J by Catchment area [2]

Catchment areas	Time of the research	Subject	Excluded	Completing interview	Response rate (%)
Nagasaki Pref. (Nagasaki City)	2002–2003	800	13	208	26.4
Okayama Pref. (Okayama City, Tamano City)	2002–2004	2,308	291	1,274	62.9
Kagoshima Pref. (Kushikino City, Fukiage, Ichiki, Higashi-ichiki)	2002–2004	1,473	114	955	70.3
Tochigi Pref. (Sano City)	2003–2004	1,051	155	550	62.1
Yamagata Pref. (Tendo City, Kaminoyama City)	2004–2006	1,684	168	770	50.8
Kanagawa Pref. (Yokohama City)	2005–2006	1,010	87	377	40.8
Total	2002–2006	8,316	818	4,134	55.1

Table 6.2 Prevalence of mood disorders (including depression) (ICD-10F, %) [2]

Age	No. of subjects	Lifetime prevalence rate	Annual prevalence rate	Monthly prevalence rate
20–34	683 (16.5)	10.4	4.8	0.4
35–44	617 (14.9)	8.1	2.1	0.3
45–54	785 (19.0)	9.7	3.6	1.0
55–64	844 (20.4)	7.9	2.4	0.7
65 and up	1,205 (29.1)	3.8	0.8	0.2
Total	4,134 (100)	7.5	2.5	0.5
Male	1,871	4.6	1.4	0.3
Female	2,263	9.9	3.4	0.7

If the frequency in younger groups hereafter remains at the same level with advancing age, this high prevalence rate in younger groups may imply an elevated future frequency in older age groups in comparison with current data. As a result, overall frequency of mood disorders will increase, that is, depression will presumably become more common. However, the reason for the lowest prevalence rate being in older groups is not easy to guess. This group may have had a lower frequency of mood disorders several decades ago when they were young, or they eliminated their own prior history of mood disorders due to reduced retrospective memory performance, or only a small number of patients in this group who had mood disorder at a young age are still alive because onset at a young age typically involves poor life prognosis, such as completed suicide. Whatever the case, the decisive factor still remains unclear.

Next, the mood disorders that were found to have a 7.5 % life prevalence rate in the subjects were categorized into subtypes in accordance with ICD-10F. As shown in Fig. 6.1, 6.63 % of these had a mild, moderate, or severe depressive episode (F32.0-3), and 0.60 % of the manic episodes included hypomania (F30.0-2) and

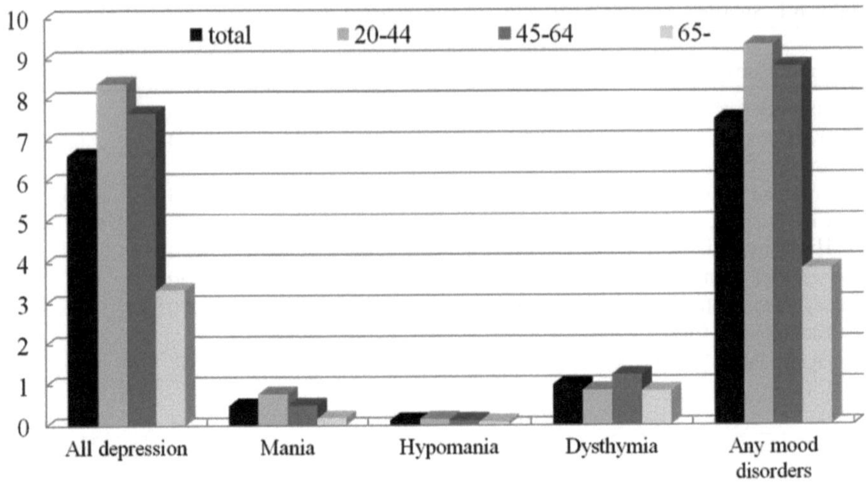

Fig. 6.1 Life-time prevalence of the mood disorders compared by subtype in each age group (ICD-10F) [2]

Table 6.3 ICD-10F lifetime prevalence rate and consulting behavior by illness (%) [2]

	Any mood disorder	Any neurotic disorder	Any disorder due to psychoactive substance	Any psychiatric disorder
Lifetime prevalence rate	7.5	9.2	4.1	17.0
Psychiatrist	20.3	15.4	16.0	14.8
General physician	13.9	16.8	7.1	13.1
Doctor total	30.3	26.7	20.7	24.3
Other expert	8.7	6.8	3.6	5.5
Other consultant	8.4	9.2	3.6	6.0
Consultation total	36.8	33.8	22.5	30.0

0.99 % dysthymia (F34.1) [2]. No significant difference was seen in frequency distribution by age group. Hypomania and dysthymia did not show up more frequently in younger age groups. This means that the recent controversy that suggests a relatively high frequency of atypical depression in the younger generation could not be verified. Despite the well-structured mood-disorder-related questionnaire in CIDI, which was comparable to normal open-ended interviews, this controversial suggestion was shown to be difficult to prove credibly by an epidemiological survey such as this one using a common interview method.

Another topic revealed by the WHO/WMH-J was the treatment pattern of the patients diagnosed with mood disorders. Table 6.3 shows the lifetime prevalence rates and treatment patterns of not only the mood disorder patients involved in this survey, but also patients with any neurotic disorders, as well as those with any mental disorders including other mental disorders than those examined in this study. As this table suggests, only about 30 % of those affected by any mood disorders underwent medical care, and the rest remained untreated [2]. As a consequence, the association between patients who were untreated and attempted/completed suicide is considered. The USA, one of the participating countries in WHO/WMH, is known for its high prevalence rate of mood disorders over a long time and showed a more than two times higher prevalence rate than that in Japan in the survey results (the annual prevalence rates in Japan and the USA was 3.1 and 9.6, respectively), but had less than half the suicide rate in the general population of that in Japan. (The completed suicide rate per 100,000 in Japan and the USA was 24.2 and 11.0, respectively.) This may be considered to imply that a thinner linkage between mood disorder patients and medical care in Japan has triggered such a high suicide rate.

6.2 Summary of the Study on the Standardized Assessment of Depressive Disorder

Table 4.2 in Part II (Chap. 4) shows the most recent mood disorder studies. Analysis of these results is still ongoing; therefore, only a part of the overall results has been summarized so far.

WHO collaborative studies have historically included psychosocial and epidemiological studies to explore the characteristics of mental illnesses in the participating regions or countries. However, with respect to mood disorders, biological studies such as a study on the use of the dexamethasone suppression test (DST) in the diagnosis of depression or a controlled study on administration routes of antidepressant drugs were also included. This is based on the financial capacity of the Division of Mental Health of WHO to conduct relevant studies where they had to rely on financial aid from external institutions that might have called for the establishment of biological findings as a prerequisite to the funding. Such biological studies are not intended to be shown here. We will focus on clinically based, psychosocial epidemiological findings as we have described for schizophrenia.

The Standardized Assessment of Depressive Disorder (SADD) study demonstrated that identification and evaluation of depressive symptoms is possible based on standardized criteria in schizophrenia studies [3]. An in-depth view of symptoms showed slight regional gaps between the participating countries.

SADD has developed its own semi-structured interview schedule for symptom assessment, known as the SADD assessment scale [4]. The frequency of depressive symptoms in depression patients obtained using this interview schedule showed distributions patterns that varied from center to center. Table 6.4 lists depressive symptoms exhibited in more than 75 % of the subjects, suggesting a regional difference between Nagasaki and Tokyo. Across all regions, known major symptoms of depression, such as *sadness, joylessness, lack of energy, and loss of ability to concentrate*, were clearly identified. There was, however, noticeably much less variety of symptoms expressed by the patients, particularly in Tokyo in comparison with

Table 6.4 Comparison of depressive symptoms observed in all participating centers in Nagasaki and Tokyo (SADD study; symptoms observed in more than 76 % patients with depression) [6]

Nagasaki	Tokyo	All centers
Sadness	Sadness	Sadness
Joylessness	Joylessness	Joylessness
Hopelessness	Anxiety	Anxiety
Anxiety	Lack of energy	Tension
Tension	Loss of ability to concentrate	Lack of energy
Lack of energy		Loss of interest
Disruption of social functioning		Loss of ability to concentrate
Slowness and retardation of thought		Feeling of incompetence, maladjustment, and lack of self-worth
Indecisiveness		
Loss of interest		
Loss of ability to concentrate		
Feeling of incompetence, maladjustment, and lack of self-worth		
Psychomotor retardation		
Inability to fall asleep		
Fitful, restless sleep		
Feelings of pressure and pain		

those in Nagasaki. Depressive patients in Nagasaki had more diversified symptoms which were considered as typical expression of depression than those in other regions [5].

Since our study on SADD was initiated almost three decades ago and is already completed, there might be some differences in results compared with those obtained in recent studies. Considering the recent patterns of atypical expression of depressive symptoms, the study method should be modified to review the results anew through the continuous implementation of studies.

Also in this SADD, the subjects were followed up for 10 years from the time point of inclusion in the study, and I engaged in this follow-up study. The included cohort were not those with first episode of depression but those with depression under treatment in collaborative centers and other cooperative medical institutions who met the inclusion criteria initially defined for symptoms. The results were narrowed to show only long-term course and outcome in a cohort with mixed depression, not limited to first episode of depression as in the DOSMeD study. In addition, comparative studies were conducted in the study on SADD by categorizing depression into two major subtypes—endogenous depression (END, including periodic depression, involutional depression, and bipolar depression) and psychogenic depression (PSY, including exhaustion depression, neurotic depression, and reactive depression)—which are not in use today but were common at that time.

The included subjects accounted for 554 at all the five participating centers combined, inclusive of 108 subjects from Nagasaki; 115 of all combined (20.8 %) and 29 in Nagasaki (26.9 %) were not available for the 10-year follow-up period.

Table 6.5 Clinical presentations of depression by subtype and social function outcomes 10 years after onset (%) [7]

	Nagasaki centers			All centers		
Outcome	END	PSY	Total	END	PSY	Total
Very good	34	28	33	42	16	39
Moderate-1	25	29	25	24	38	26
Moderate-2	10	0	9	16	25	17
Extremely poor	31	43	32	18	21	18
Ten-year outcomes in terms of suicide attempt (%)						
Attempted suicide	7	43	11	10	30	14
Completed suicide	7	14	14	10	14	11

Note: *Moderate-1* no hospitalization during the follow-up, *Moderate-2* at least one rehospitalization during the follow-up

As shown in Table 6.5, 39 % of all combined and 33 % in Nagasaki showed a very good outcome, while 18 % and 32 % showed a very poor outcome in the 10-year follow-up results, respectively. Outcomes obtained in Nagasaki were distributed in a ratio of about one-third, showing a rather poor outcome pattern compared to other centers.

Viewing this by subtypes, more PSY cases were associated with poor outcomes than END cases.

With respect to the suicide rate, 11 % of all combined and 14 % in Nagasaki completed suicide, and an apparently higher rate of attempted suicide was seen in patients with PSY than those with END [7].

Similarly, we garnered support of the WHO Western Pacific Regional Office (WPRO) to conduct a multinational comparative study entitled *The Comparative Study of Affective Disorders in Three Asian Countries* in collaboration with Shanghai Mental Health Center in Shanghai, China, and Yonsei University in Seoul, South Korea.

The initiation of this study was triggered by an indication that countries in Asia have lower prevalence rates of depressive disorder in comparison with western countries, and, particularly, China seems to have no depression patients. In addition, diverse treatment patterns of depression patients in Asia were also pointed out. The implication persuaded me of the need to at least show plausible evidence from Asia.

Our study results demonstrated that, despite the different historical trends in the concept of depression in each country, the frequency of depression was not significantly different from that seen in western countries, but, rather, differences in factors such as recognition, expression, and treatment patterns of the patients; differences in pathologic recognition of pathology of depressive symptoms by providers of mental health services and the interview schedules used; and differences in concepts of the disorder were reflected in the gaps in results. The detailed study report is published in *Acta Psychiatrica Scandinavica*, 78 [8] and 84 [9]. Thus, additional reference is omitted here.

References

1. Kessler RC, Üstün TB. The World Mental Health (WMH) Survey Initiative Version of the World Health Organization (WHO) Composite International Diagnostic Interview (CIDI). Int J Methods Psychiatr Res. 2004;13:93–121.
2. Ono Y, Kawakami N, Nakane Y, et al. Prevalence of and risk factors for suicide-related outcomes in the World Health Organization World Mental Health Surveys Japan. Psychiatry Clin Neurosci. 2008;62:442–9.
3. Sartorius N, Davidian H, Ernberg G, et al. Depressive disorders in different cultures; report on the WHO collaborative study on standardized assessment of depressive disorders. Geneva: WHO; 1983.
4. Hirota N. Studies on Standardized Assessment of Depressive Disorder. In: Committee for celebration of Professor R. Takahashi's transfer to the Dept. of Psychiatry at Tokyo Medical and dental University, editor. The Second Journal of the Nagasaki Collaborating Center of WHO for Research in Functional Psychoses – The History of the Department of Neuropsychiatry Nagasaki University School of Medicine from 1967 to 1984 – In commemoration of the achievement of Prof. R. Takahashi in our department. Nagasaki: Department of Neuropsychiatry Nagasaki University School of Medicine; 1985. p. 142–50.
5. Takahashi R, Fujii I, Nakane Y. Manic-depressive psychosis. In: Shinfuku N, Shimazono Y, editors. Psychiatry, vol. 2. Tokyo: Kanahara-Shuppan; 1983. p. 377–442 (in Japanese).
6. Takahashi R, Nakane Y. International comparative study of depressive disorder – socio-cultural and epidemiological topics. In: Okuma T, editor. Clinics and theories of depression. Tokyo: Igaku-Shoin; 1990. p. 355–75 (in Japanese).
7. Thornicroft G, Sartorius N. The course and outcome of depression in different cultures: 10-year follow-up of the WHO collaborative study on the assessment of depressive disorders. Psychol Med. 1993;23:1023–32.
8. Nakane Y, Ohta Y, Uchino J, et al. Comparative study of affective disorders in three Asian countries. I. Differences in diagnostic classification. Acta Psychiatr Scand. 1988;78:698–705.
9. Nakane Y, Ohta Y, Radford M, et al. Comparative study of affective disorders in three Asian countries. II. Differences in prevalence rates and symptom presentation. Acta Psychiatr Scand. 1991;84:313–9.

Chapter 7
Studies on Patient with Affective Disorders in General Practice

As the results in the previous chapter indicated, patients with psychiatric problems do not always go to see a psychiatrist but most likely go to be seen a general physician or primary care physician. It is pointed out that these days the number of such patients is increasing.

Professor David Goldberg published an interesting report based on numerous regional studies in the UK, including an analysis of the association between physical disorders and psychiatric disorders in outpatients visiting a primary care (or general care) clinic [1].

He reported findings, which may be a little bit old: 45.6 % of patients visited a physician with only a physical disorder, 16.8 % visited for a consultation on a child's disease, 8.3 % seemed slightly nervous but presented with any physical disorder, and the other 23.4 % presented with a mental problem. Table 7.1 shows the details.

Goldberg reported that fewer than 10 % of patients visiting a physician were considered to have no physical symptoms but to have a mental disorder and that about 15 % of patients would be definitely diagnosed as having a mental disorder associated with some physical condition.

In addition, he thought the process that the patient population with common mental-health-related problems took to visit a psychiatrist went as follows. He called it "Five Levels and Four Filters" and explained it using the models shown in Fig. 7.1.

According to Goldberg, Level 1 is the community level, and about 300 per 1,000 people a year are living with some psychiatric problem in the community. In Level 2, about 80 %, or 230 of those patients, visit a medical institution, such as a primary care clinic. In Level 3, psychological problems are observed by a physician in less than half, or about 100 of those who presented to a primary care institution. In Level 4, three-fourths of the patients are treated by a primary physician, and the other one-fourth is referred to a psychiatrist. And in Level 5, one-fifth of the referred patients are psychiatrically hospitalized. He stated that each level has a filter to go though prior to the next level, and ease of approach to psychiatric care is dependent on the transparency of the filter. He indicated that psychiatrists examined no more than 10 % of patients living with mental problems in the community.

Y. Nakane, *Progress in Social Psychiatry in Japan: An Approach to Psychiatric Epidemiology*, DOI 10.1007/978-4-431-54103-5_7, © Springer Japan 2012

Table 7.1 Association between physical disorders and psychiatric disorders in primary care setting [1]

1. Physical disorders secondarily associated with psychiatric disorders: 1.8%
   In this case:
   (a) All physical symptoms are attributable to the physical disorder
   (b) The physical symptoms are not reduced by treatment of the psychiatric disorder
   (c) The psychiatric disorder would not develop without the physical disorder
2. Physical disorders accidentally associated with unrelated psychiatric disorders: 5.4%
   In this case, there is no etiological association between both disorders, and treatment for one disorder would not affect the other disorder
3. Psychiatric disorder developing physical symptoms: 9.4%
   In this case:
   (a) Despite obvious psychiatric disorders (identified with the international classification)
   (b) Patients think that all the problems are attributable to physical disorders presenting physical symptoms or that the manifested symptoms might induce some mental problem, thus,
   (c) Psychiatric consultation for the physical symptoms will be necessary, and,
   (d) Only treatments of psychiatric disorders can resolve or reduce the physical symptoms
4. Psychiatric disorders diagnosed completely: 7.8%
   In this case:
   Patients have no physical symptoms or they seldom conceive of them as a part of psychiatric disorders

Five levels	Four filters	Prevalence
Level 1 (community)		260-315/1,000/year
	First filter (Disease behavior)	
Level 2 (patients receiving primary care –total number of patients with psychiatric disorder)		230/1,000/year
	Second filter (Power detecting disorders)	
Level 3 (number of patients diagnosed to have psychiatric disorders by primary care physicians)		101.5/1,000/year
	Third filter(Refer topsychiatric care/service)	
Level 4 (number of patients receiving psychiatric care service)		23.5/1,000/year
	Fourth filter (Psychiatric admission)	
Level 5 (number of inpatients in psychiatric wards)		5.71/1,000/year

Fig. 7.1 Five levels and four filters for common mental disorders [1]

As a result, the diagnostic power of primary care physicians and general physicians becomes important. It would be reasonable for primary care physicians to treat patients with less severe suffering if they can evaluate such suffering correctly and provide appropriate treatments for them.

The Nagasaki Collaborating Center participated in two relevant collaborative cross-national studies: *Studies on Depressive Patients in General Practice (GP study), 1977–1980* [2] and the *WHO Collaborative Study on Psychological Problems in General Health Care Settings (PPGHC), 1990–1996* [3].

7.1 Studies on Referral Rates in Patients with Depression in the General Health Care Setting

The Nagasaki Collaborating Center conducted the GP study focusing only on depression in tandem with the aforementioned SADD study. First, a physician at the clinic examined a patient selected from four to five patients who visited either the internal medicine clinic of the national general hospital or the public general hospital in Nagasaki for the first time and held an interview on psychiatric problems based on the GP screening sheet. Second, a psychiatrist held another interview with the patients based on the depression criteria and based on the SADD assessment scale independently from the physician. Finally, the frequency of differences in diagnosis made by physicians and psychiatrists was determined after the psychiatrist made a diagnosis according to ICD-9.

The screening sheet was prepared for physicians prior to the research. Physicians were to check whether a patient had a physical problem only or a physical problem associated with a psychological problem and state their opinion as to whether the patient's psychological problem was *mainly anxiety*, *mainly depression*, or *other problem*. In turn, psychiatrists used inclusion criteria consisting of eight depressive features (depressed mood, suicidal ideation, feeling of despair, feeling of worthlessness, hypochondriasis and anxiety, decreased vitality and performance, feeling remorse, and feeling lack of joy) in the interview. Patients with at least two depressive features were included in the study. The selected patients underwent a structured interview based on the SADD assessment scale. Physicians and psychiatrists examined new outpatients independently and compared the diagnoses; when 50 outpatients with depression were selected by the psychiatrists, the screening was completed.

Consequently, as shown in Table 7.2, a total of 831 new outpatients were screened to obtain 50 patients with depression, which accounted for 6.0 % of new outpatients who visited the internal medicine clinics. Through the process of this selection and inclusion, physicians were able to exclude 99.1 % of the patients who did not have depression (specificity) but recognized only 8 of the 50 patients with depression (16 %) correctly (sensitivity). This low sensitivity suggested the need to improve awareness of depression in general physicians and primary care physicians. It also suggested that the lack of recognition of depression by these physicians might be largely associated with the symptom distributions observed in internal medicine outpatient clinics. Expression of depressive symptoms was significantly different between Tokyo and Nagasaki, as previously described. Similarly, all patients who

Table 7.2 Patients enrolled in the GP study [2] (831 outpatients of internal medicine clinics)

		Diagnoses made by psychiatrists		
		Depression	Other than depression	Total
Screening by physicians	Depression	8 True positive	7 False positive	15
	Other than depression	42 False negative	774 True negative	816
	Total	50	781	831

Sensitivity = 8/50 = 16 %; specificity = 774/781 = 99.1 %

visited internal medicine outpatient clinics did not present with typical symptoms and had more physical symptoms than psychological symptoms [2].

The patients were categorized into the following subtypes: 20 of 50 patients had ICD-9/296.1, manic depression with depressive episodes only (unipolar depression); 10 had ICD-9/298.0, reactive (psychogenic) depression; 7 each had 300.4, neurotic depression, and 309.0, adjustment reaction with brief depressive reaction, respectively; and the rest had various depressive symptoms. Expression of typical depressive symptoms was thus limited. In addition, as shown in Table 7.3, the prevalence of depressive symptoms observed in the GP study was significantly lower than that in the SADD study.

7.2 Studies of Medical Care in Patients with Psychological Problems in the General Health Care Setting

Another study we were engaged in was the WHO Collaborative Study on Psychological Problems in General Health Care Settings (PPGHC) [3]. This PPGHC study was initiated in 1990, aiming to investigate how many patients who visited outpatient clinics of internal medicine had psychiatric disorders and whether those patients were able to receive appropriate treatment, and to discuss what measures psychiatrists should take for them. The follow-up to this PPGHC study was completed in 1996. It was a large-scale study conducted by 15 WHO Collaborating Centers in 13 countries to compare medical care in the general medical care setting, and it was launched in a bigger way than previous WHO studies. Photographs of the first Researchers' Meeting (July 1990) and several slides of lists, including participating Center chiefs and countries and the front cover of the initial report, are shown in Figs. 7.2, 7.3, 7.4, 7.5, and 7.6.

While the general physicians and primary care physicians were seeing more patient visits with mental health problems than in the past, and these physicians were becoming interested in psychiatric treatments, the psychiatrists were interested in what treatments the general physicians were giving to those patients. The need to collect such data was growing internationally since Professor Goldberg

Table 7.3 Positive rate of physical symptoms compared between the GP study and the SADD study (%) [1, 2]

Symptoms	GP study	SADD study	Symptoms	GP study	SADD study
Sadness	91	93	Fitful, restless sleep*	66	82
Joylessness***	68	88	Lack of appetite+	66	79
Hopelessness***	42	82	Change of body weight	58	65
Anxiety/tension*	74	86	Constipation	39	49
Irritated feeling	45	46	Feelings of pressure/pain	68	73
Lack of energy*	80	89	Other physical symptoms***	76	46
Disruption of social functioning***	59	86	Decrease in libido**	28	52
Slowness or retardation of thought***	56	83	Change of perception of time***	43	65
Indecisiveness***	43	71	Suicidal ideas***	23	66
Ideas of insufficiency, inadequacy and worthlessness etc.***	50	87	Feeling of guilt/self-reproach**	14	39
Loss of interest***	50	84	Hypochondriasis	48	52
Loss of ability of concentration***	56	87	Idea of impoverishment*	15	33
Early awakening*	45	65	Ideas of persecution/self-reference**	7	30
Inability to fall asleep+	63	77	Psychomotor retardation***	54	83
			Psychomotor agitation***	9	47
			Worse in the morning*	33	53
			Worse in the evening	12	14

GP study: $n=50$; SADD study: $n=108$
*$p<0.05$; **$p<0.01$; ***$p<0.001$; +$p<0.1$

Fig. 7.2 A PPGHC researchers' meeting (1990.7.2–7.6) – at WHO/HQ

Fig. 7.3 A PPGHC researchers' meeting (1990.7) – at WHO/HQ

et al. issued the aforementioned indications. It was a rather major challenge for the then Division of Mental Health of WHO. Most of the WHO regional research centers that had ever participated in international collaborative studies participated in this study.

We were engaged in this study more energetically than before, with anticipation of obtaining the first findings in Japan. In addition, we also anticipated compatibility of findings in Asia due to the participation of the Shanghai Collaborating Center.

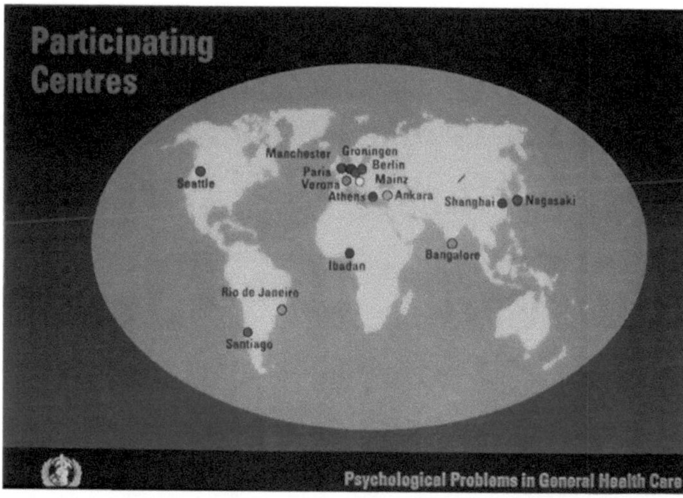

Fig. 7.4 PPGHC participating centers on the world map

Fig. 7.5 PPGHC participating center chiefs and countries

The Nagasaki Collaborating Center conducted the study at the same internal medicine clinics of the national general hospital and the public general hospital that participated in the previous study and asked for the cooperation of the physicians. Considering the daily medical care in both hospitals and the manpower of the Center, we set the study duration for half a year from March to the end of August 1991 and included return patients of 15–65 years of age who agreed to cooperate with the study.

Fig. 7.6 A PPGHC report

The evaluation sheet used in the first study phase was the General Health Questionnaire, 12-item version (GHQ-12) [1], which was developed by Professor Goldberg and was considered to be useful to detect and determine symptoms of patients with minor psychiatric symptoms. Although GHQ-60 was already translated into Japanese and published by the National Institute of Mental Health by 1981 [4], we did the translation and reverse translation anew to prepare GHQ-12 for this study.

Professor Goldberg had already predicted an association between the GHQ-12 score and psychological disorders, as shown in Table 7.4: the low score arm of 0 or 1 point would account for 60 % of the total patients, the middle score arm of 2 or 3 points, and the high score arm of more than 4 points would account for 20 % each, respectively; and the incidence rate of psychological disorders would be almost even in each score arm. Therefore, patients were selected for the second-stage interview with due consideration of balanced selection from each arm. For this, the selection rate was set by score: 100 % of the patients in the high score arm, 35 % from the middle score arm, and 10 % from the low score arm were selected for the second-stage interview. The patients from the middle score arm and the low score arm were randomly selected. We explained the gist and contents of this study to physicians of both hospitals and asked them to participate in the study collaboration as physicians for the physical and mental evaluations of the patients. Candidate patients for the second-stage interview were given an explanation about the study

Table 7.4 Association between the GHQ-12 score and frequencies of psychological impairment (based on assumption that the 1,000 patients were tested with GHQ-12) [1]

GHQ-12	Low score arm (0–1)	Middle score arm (2–3)	High score arm (4–)	Total
Patients with mild disorders or healthy population (75 %)	540	170	40	750
Patients with borderline psychological disorder (15 %)	45	15	90	150
Patients with evident psychological disorder (10 %)	15	15	70	100
Total	600	200	200	1,000

Table 7.5 Diversified/multi-method evaluation in the PPGHC study [1]

	Psychiatric aspects	Physical/medical aspects	Dysfunctional aspects
Patients	The General Health Questionnaire (GHQ-12 and -28)	Complaints about general health	Simple dysfunction questionnaire Number of days with dysfunction
Physicians	Severity evaluation of psychiatric disorders Psychiatric diagnosis	Evaluation of general health Internal medicine assessment Evaluation of physiomedical severity	
Psychiatrists	Composite International Diagnostic Interview (CIDI) Psychiatric diagnosis Evaluation of association between psychological disorders and physical symptoms	Evaluation of chronic diseases	The Groningen Social Disability Schedule (GSDS)

and were enrolled in the second-stage interview when their consent and permission were obtained.

The second-stage interview was conducted by psychiatrists based on the primary care version of the composite international diagnostic interview (CIDI/PC), which was developed by WHO [5]. The semi-structured interview was also conducted based on the Groningen social disability schedule (GSDS), which was developed by Wiersma and Ormel in 1988 [6] to evaluate disorder-related social dysfunction levels. Evaluation with both scales was practiced frequently prior to these interviews for reliable evaluation.

Evaluation sheets were also prepared for the physicians who agreed to participate in this study to evaluate and describe main complaints and diagnosis, categorization of the current symptoms, comprehensive evaluation of health, severity of physical/medical disorder, severity of psychiatric disorder, and the treatments provided to the patients. The diversified evaluation processes with multiple scales are shown in Table 7.5.

Similar to procedures of the DOSMeD study, we conducted the pilot study at the internal medicine clinics of both general hospitals prior to the main study and examined the feasibility and potential problems of the study. In particular, we tried sampling for the first-stage interview, screening with GHQ-12, and the setting of the second-stage interview several times. We asked the patients who underwent the second-stage interview to receive 3-month and 1-year follow-ups.

A total of 1,580 patients were selected from 2 internal medicine outpatient clinics, and a total of 1,555 patients consisting of 759 male patients and 796 female patients agreed to participate in the main study and completed the GHQ-12. Of the 1,555 patients, 1,043 patients and 512 patients were enrolled in the study at the national hospital and the public hospital, respectively. The difference in the number of patients was due to the difference in the number of physicians who participated in the study as well as the study duration.

A total of 483 patients consisting of 104 from the GHQ-12 low score arm, 92 from the middle score arm, and all 287 patients from the high score arm were selected for the second-stage interview. Most of the patients participated in the second-stage interview on the same day as the first-stage interview, and a very few unavailable patients were requested to visit the clinic again within several days for the interview. Consequently, a total of 336 patients (69.6 %), consisting of 81 from the low score arm, 66 from the middle score arm, and 189 from the high score arm completed the second-stage interview. Table 7.6 shows the number of patients participating in the scheduled two follow-ups.

Table 7.7 shows the results of the second-stage interview in patients who were selected based on the first-stage interview results.

Of 336 patients who participated in the second-stage interview, half or fewer patients in the GHQ low score arm were aware that they were not in good general health. However, the awareness increased in the middle and high score arms as the score was increased, and 83 % of patients in the middle score arm and 88 % in the high score arm were aware of their poor general health. Along with the increased awareness, the prevalence of decreased physical functions also increased in both higher score arms. Assessment of the patients by physicians was generally lower than patients' self-awareness and slightly higher than 10 % of their self-evaluations. Significantly, prevalence of poor physical health and physical disorder assessed by physicians was highest in the GHQ middle score arm. While the incidence of psychiatric problems was insignificant in patients in the low score arm, it was remarkably higher in patients of the middle and high score arm.

However, the frequency was no more than one-third of the problems assessed by psychiatrists.

Psychiatric evaluations suggested abnormality conforming to the GHQ score in patients and also that the study using the GHQ method was reasonable: while a clinical psychiatric diagnosis was made in about 7 % of patients with the GHQ low score, psychiatric diagnosis increased to 30 % and 50 % in both middle and high score arms, respectively, as the score increased. Similar results were obtained with the CIDI. The incidence of social dysfunctions was about double that for psychiatric diagnosis, which indicated that we should provide further social function support to

Table 7.6 Longitudinal transition of patients in the PPGHC study [1]

Sampling	One selected from 4 to 5 return patients who visited internal medicine outpatient clinics during the study period; 1,580			
GHQ-12 score	Low score arm	Middle score arm	High score arm	Total, *N*
Research population	1,001 (64.4 %)	267 (17.2 %)	287 (18.5 %)	1,555
Screening for the second-stage interview	10 % random sampling	35 % random sampling	100 % sampling	
	104 (21.5 %)	92 (19.0 %)	287 (59.4 %)	483
Patient's agreement and physician's permission for the second-stage interview				
Patients receiving second-stage interview	81 (24.1 %)	66 (19.6 %)	189 (56.3 %)	336
Patient's agreement and physician's permission for follow-ups				
3 months follow-up	74 (25.5 %)	58 (28 %)	158 (54.5 %)	290
Patient's agreement and physician's permission for follow-ups				
1 year follow-up	49 (27.4 %)	36 (20.1 %)	94 (52.5 %)	179

Table 7.7 Association between 3-layered GHQ score and main evaluation (%) [1]

	Low score arm	Middle score arm	High score arm	Overall
Patients of second-stage interview	81	66	189	336
Self-evaluation of patients				
General health	49.4	83.3	88.4	78.0
Physical dysfunction	21.0	30.8	52.7	40.5
Evaluation by physicians				
General health	5.2	12.5	10.3	9.5
Severity of psychiatric disorder	1.6	10.9	12.8	10.1
Physical/medical severity	48.7	65.6	54.1	55.1
Evaluation by psychiatrists				
Psychiatric clinical diagnosis	7.4	33.3	52.4	37.8
CIDI assessment	11.1	30.3	45.5	34.2
GSDS	27.2	48.5	67.2	53.9

Note: *Self-evaluation by subject*: the frequencies of "not too bad" or "not good" were shown out of "excellent," "very good," "good," "not too bad," and "not good" as self-evaluation for "general health conditions." The frequencies of "mild disorder" or higher severities were shown out of "no disorder," "mild disorder," "moderate disorder," and "severe disorder" as self-evaluation for "physical dysfunction"; *Evaluation by physician*: the frequency of "poor" was shown out of "excellent," "very good," "good," "usual," and "poor" as the evaluation for "general health conditions." The frequencies of "mild disorder" or higher severities were shown out of "normal," "have some symptoms leading to an illness," "mild disorder," "moderate disorder," and "severe disorder" as severity of "psychiatric disorder." The frequencies of "mild" or higher severities were shown as evaluation of "physical/medical severity"; *Evaluation by psychiatrist*: the frequency of "present" was shown as evaluation for "psychiatric clinical diagnosis" and CIDI assessment. The GSDS evaluation showed "mild disorder" or higher severities of overall assessments of "no disorder," "mild disorder," "moderate disorder," and "severe disorder"

Table 7.8 Psychiatric disorders detected through the CIDI in the PPGHC study and compared by sex and age [1]

ICD-10F	Prevalence in all patients	Male	Female	15–24 years of age	25–44	45–65
F10.1/10.2						
Alcohol-related problems	6.2	11.5	1.2	7.0	7.2	5.8
F32/33 Currently depressed	2.6	2.3	2.8	12.3	2.3	2.1
F34 Dysthymia	0.4	0.4	0.4	2.5	0	0.4
F41.0 Panic disorder	0.2	0.2	0.2	0	0.4	0.1
F41.1 Generalized anxiety disorder	5.0	6.5	3.6	0	2.6	6.2
F45.0 Somatoform disorder	0.1	0	0.2	0	0	0.1
F48.0 Neurasthenia	3.4	3.8	3.0	2.2	6.6	2.3
Some psychiatric disorders	14.8	19.9	9.9	21.8	14.7	14.4
At least two psychiatric disorders	2.8	4.0	1.6	2.2	3.0	12.7

patients who visit an internal medicine outpatient clinic regularly but are not well aware of their social dysfunction.

Table 7.8 shows the prevalence of psychiatric disorders in the 1,555 patients who visited an internal medicine outpatient clinic, which was estimated from results of the second-stage interview in 336 patients.

Patients who presented to the clinic with F10, alcohol-related troubles, were unexpectedly the most common: 3.7 % with F10.2, alcohol dependence; 2.5 % with F10.1, alcohol abuse; and 6.2 % combined prevalence were observed in both male and female patients. The table shows that the second most common disorder was F4, neurotic disorders; however, depressive disorders might be the most common psychiatric disorder, with higher incidence than alcohol-related disorders, as F48.0, neurasthenia, may be included in depressive disorders.

The incidence rate of 6.4 % was almost similar to that of 6.0 % in the GP study, which had been conducted about 10 years before this study with collaboration of physicians in the same two hospitals. This result denies the common belief that the number of patients with depression who visit general health care or internal medicine outpatient clinics is markedly increasing. We were able to obtain the finding that the number of female patients with depression was not significantly greater than the number of male patients, and that multiple diagnoses were likely to be made in many male patients. More patients with depression-related disorders were observed in the young population than the aged population. This is linked to the recent findings of WHO/WMH-J.

As seen in Table 7.9, it is unfortunate that there was a big difference between physicians' awareness of psychiatric disorders and psychiatrists' assessments. More than ten patients who were diagnosed by psychiatrists to have disorders of the ICD-10F category were included in F41.1, generalized anxiety disorder; F32/33, currently depressed; F48.0, neurasthenia; and F10, alcohol-related problems. In addition, generalized anxiety disorder, currently depressed, and neurasthenia were common in order of decreasing number of patients.

Table 7.9 Physicians' awareness of patients with psychiatric disorders observed in the PPGHC study [1]

		N of patients perceived by physicians		
ICD-10F	*N* of all	*n*	%	Estimated awareness
F10.1 Alcohol dependence	18	0	0.0	0.0
F10.2 Alcohol abuse	16	1	6.3	3.6
F32/33 Currently depressed	25	5	20.0	19.3
F34 Dysthymia	4	1	25.0	26.4
F41.0 Panic disorder	2	0	0.0	0.0
F41.1 Generalized anxiety disorder	30	9	30.0	22.5
F45.0 Somatoform disorder	1	0	0.0	0.0
F45.2 Hypochondriasis	3	0	0.0	0.0
F48.0 Neurasthenia	22	4	18.2	10.9
Some psychiatric disorders	62	15	24.2	18.3
At least two psychiatric disorders	20	3	15.0	13.1
Total	*336*	*31*	*9.2*	*4.8*

Of patients with generalized anxiety disorder or depression, 25 patients with depression were diagnosed by physicians to have gastrointestinal disease (10), malignant tumor (4), cardiovascular disease (4), endocrine/metabolic disease (2), neuropsychiatric disorder (2), respiratory disease (1), symptoms without abnormal clinical test results, and unknown symptoms (1). On the other hand, 25 patients with generalized anxiety disorder were diagnosed to have cardiovascular disease (6), gastrointestinal disease (5), respiratory disease (4), endocrine/metabolic disease (2), malignant tumor (1), skin disease (1), symptoms without abnormal clinical test results (1), and unknown symptoms (1). It is not denied that these disorders or conditions might underlie respective psychiatric disorders and that deeper understanding could be possible if physicians were more interested in the psychological aspects of those patients.

A total of 25,916 patients, including 1,555 patients of the Nagasaki Collaborating Center, were enrolled in the study by 15 collaborating centers. Of those patients, 24 % of patients in all centers who visited general health-care clinics were diagnosed to have some psychiatric disorder of ICD-10F, as shown in Table 7.10. Alcohol-related problems included alcohol dependence and alcohol abuse; depression also included dysthymia in this table. Neurasthenia is a term for neurotic disorder but is clinically similar to depressive state, as described above. Generalized anxiety disorder was traditionally referred to as anxiety neurosis. The incidence of panic disorder was shown at the bottom of the table.

This table shows diagnostic rates of the Collaborating Centers in Manchester, Seattle, and Nagasaki for comparison. The rates other than those for alcohol problems were relatively lower in the Nagasaki Collaborating Center. These are the results of diagnoses that physicians and psychiatrists made independently for the same patients with a specific scale. As mentioned frequently, it is evident that about 15 % of patients who saw a general physician or visited an internal medicine outpatient

Table 7.10 Psychiatric diagnosis in patients who visited a primary care physician in the PPGHC study (%) [1]

ICD-10F diagnosis	All centers	Manchester	Seattle	Nagasaki
Alcohol-related problems	3.7	2.3	1.5	6.2
Depression	10.5	16.9	6.3	3.0
Neurasthenia	5.5	9.7	2.1	3.4
Generalized anxiety disorder	7.9	7.1	1.9	5.0
Panic disorder	1.1	3.5	20.4	0.2
Patients with some diagnosis	24.0	26.2	20.4	14.8

Table 7.11 Primary care physicians' awareness of psychiatric diagnosis (%)

ICD-10F diagnosis	Manchester	Seattle	Nagasaki
Alcohol-related problems	66.1	44.3	0.0
Depression	69.6	56.7	19.3
Neurasthenia	49.8	46.8	10.9
Generalized anxiety disorder	72.3	31.9	22.5
Panic disorder	70.6	76.9	0.0
Patients with some diagnosis	62.9	56.9	18.3

clinic for treatment had psychiatric problems. Although the most common problem in those patients was depressive disorders, awareness of patients with psychiatric problems in physicians was low, and the patients were diagnosed to have some disease of internal medicine.

Please see Table 7.11 based on Table 7.10. Physicians commonly made similar diagnoses to psychiatrists in both the Manchester and Seattle Collaborating Centers. Particularly, nearly 70 % of physicians were aware of most diagnoses of ICD-10F in the Manchester Center. The awareness was lower in the Seattle Center than in the Manchester Center, but much higher than in the Nagasaki Center. These findings may indicate that the system of primary care physicians or family physicians has already been established in these areas and poses the question of how much excessively pursuing specialties in medical care in Japan and other areas can answer community needs.

WHO's purpose is the development of primary care physicians who can make psychiatric diagnoses. To achieve this purpose, WHO is developing (1) texts for diagnoses and treatment of psychiatric disorders and (2) texts to educate patients and their families. The Nagasaki Collaborating Center has worked together to publish the Japanese versions.

These findings also provided evidence suggesting the need to study *appropriate response to common mental diseases* as a common practice for a general physician rather than a topic for a proper psychiatrist through the mandated intern program after graduation from a medical school.

References

1. Goldberg D, Huxley P. Common mental disorders – a bio-social model. London/New York: Tavistock/Routledge; 1992 (translated into Japanese by Nakane Y, Sozo-shupan, Tokyo, 2000).
2. Fujii I. Studies on depressive patients in general practice. In: Committee for celebration of Professor R. Takahashi's transfer to the Dept. of Psychiatry at Tokyo Medical and dental University, editor. The Second Journal of the Nagasaki Collaborating Center of WHO for Research in Functional Psychoses – The History of the Department of Neuropsychiatry Nagasaki University School of Medicine from 1967 to 1984 – in commemoration of the achievement of Prof. R. Takahashi in our department. Nagasaki: Department of Neuropsychiatry Nagasaki University School of Medicine; 1985. p. 150–60.
3. Nakane Y, Michitsuji S. Results from the Nagasaki centre. In: Üstün TB, Sartorius N, editors. Mental illness in general health care – an international study. Chichester: Wiley; 1995. p. 193–210.
4. Goldberg DP. The detection of psychiatric illness by questionnaire. A technique for the identification and assessment of non-psychotic psychiatric illness. Maudsley Monographs No. 21. London: Oxford University Press; 1972 [translated into Japanese by Nakagawa Y, Ohbou I (1981), National Institute of Mental Health Japan].
5. Division of Mental Health, World Health Organization. The composite international diagnostic interview, core version 1.1. Washington: APP; 1996.
6. Wiersma D, Ormel DA. The Groningen social disabilities schedule: development, relationship with I.C.I.D.H., and psychometric properties. Int J Rehabil Res. 1988;11:213–24.

Chapter 8
Development from Atomic Bomb Diseases to Survivors' Mental Health Support: Findings on Disaster Psychiatry

When I was promoted to university lecturer in the Department of Neuropsychiatry of Nagasaki University Hospital after graduation from graduate school and was able to have my own laboratory in one corner of the medical office ("ikyoku," see Appendix at the last part of this chapter), I started to call it the *epidemiology laboratory*. Initially, the purpose and aim of my studies had a similar vision to that of those who came before me, where epidemiology in the psychiatric field in Japan historically tended to have a strong aspect of genetic epidemiology. On the other hand, however, the genetic aspects of my epidemiological studies had to shift to social psychiatry in line with the gradual change in my belief that epidemiology should constitute a basic medical science of social psychiatry. It was like a stumble in a dark, since the accumulated data in psychiatric epidemiology in Japan up to that time were not instrumental enough to provide social psychiatry with basic information. Accordingly, the *epidemiology laboratory* found no welcome among my colleagues in the medical office; *social psychiatry laboratory* seemed instead to be more acceptable. Eventually, the name of our study group ended up being called *the social group* until we started conducting actual research activities.

Cooperation from numerous researchers is the foremost prerequisite to advanced epidemiological studies; funding for studies is the second, along with abundant and affordable time. Fortunately, from the human resource viewpoint, the medical office was able to welcome enough junior fellows who wished to go into the Department of Neuropsychiatry to be assigned to the planned studies. With respect to funding sources, however, despite repetitive applications, the chances that an epidemiological study would receive science research grants from the Ministry of Education, Science, Sports, and Culture, the Ministry of Health and Labor, or any other funds were small in comparison with biological studies that are directly linked to etiology.

We had to be well prepared to continue and complete a regional epidemiological study. Not a few studies turned out to be difficult for the responsible study group to pursue in the course of study. For some studies, in consideration of expected adversities in the preparatory stage, we had to consider the possibility of a withdrawal already in the pilot study that we conducted to objectively assess the feasibility of

Y. Nakane, *Progress in Social Psychiatry in Japan: An Approach to Psychiatric Epidemiology*, DOI 10.1007/978-4-431-54103-5_8, © Springer Japan 2012

the main study. A decision on withdrawal, however, is not always easy to make for international comparative studies and multicenter collaborative studies.

In the meantime, study systems were able to be established step by step in the university and surrounding areas after multiyear efforts, and we believe that appropriate achievement of the efforts mentioned earlier was made.

Particularly as a medical scientist living in Nagasaki, I had always been considering devoting myself to establishing potential psychiatric findings in mental health problems among atomic bomb survivors and sufferers exposed in August 1945 in Hiroshima and Nagasaki, and to develop foundations for proper support systems for them within an administrative framework, in addition to my clinical support as a psychiatrist.

As briefly discussed in Chap. 3, the Nagasaki University School of Medicine suffered devastating physical damages from the atomic bomb dropped by a B29 on a tennis court at 171, Matsuyama, Urakami area in northern Nagasaki City at 11:02 a.m. on Wednesday, August 9, 1945. Two B29s were flying at a height of 10 km and dropped an atomic bomb at a height of 9,600 m, which exploded at a height of about 500 m. On that day, it was reported to be sunny, with very hot temperatures, and little wind in Nagasaki. According to the report on the post-disaster situation by the Commission for the Conservation of Atomic-Bomb Documents (issued in July 1950) [1], there were 73,884 deaths, 74,909 injured, 120,820 afflicted (number of household members within a 4-km radius from the epicenter that were completely destructed by fire or completely collapsed), 18,409 afflicted houses (all of those within the 4-km radius of the epicenter, about 36 % of all the houses in Nagasaki City), 11,574 houses completely collapsed by fire (within a 4-km radius from the epicenter, about one-third of all the houses in Nagasaki City), 1,326 completely collapsed houses (within a 1-km radius from the epicenter), 5,509 half-collapsed houses (within a 4-km radius from the epicenter) (2009 Summary on Atomic-Bomb Survivor Relief Project, published in July 2009) [1].

Figure 8.1 shows the situation at the epicenter before and after being bombed. The Nagasaki University School of Medicine and the University Hospital are reported to be located about 550 and 700 m from the epicenter, respectively. Most of the buildings seen around the epicenter after being bombed were completely destroyed down to the vacant lot.

Despite the passage of considerable time from the atomic bombing, only an extremely small amount of research data on the psychological impacts on atomic bomb survivors were available, including those obtained in Hiroshima. Scientific research should have been implemented according to a deliberate study design. Many findings of the physical impacts on atomic bomb survivors were established, and a support system was gradually being developed, even though it was not a complete system. However, most of the views on the psychological impacts did not earn social recognition. Psychiatry focused on microcephaly in fetuses developed through exposure.

The nuclear powers did not show any sign of renouncing nuclear tests, which caused continuing global concern. About 15 years ago, France announced its resumption of nuclear testing from June 1995 against worldwide wishes and enforced the

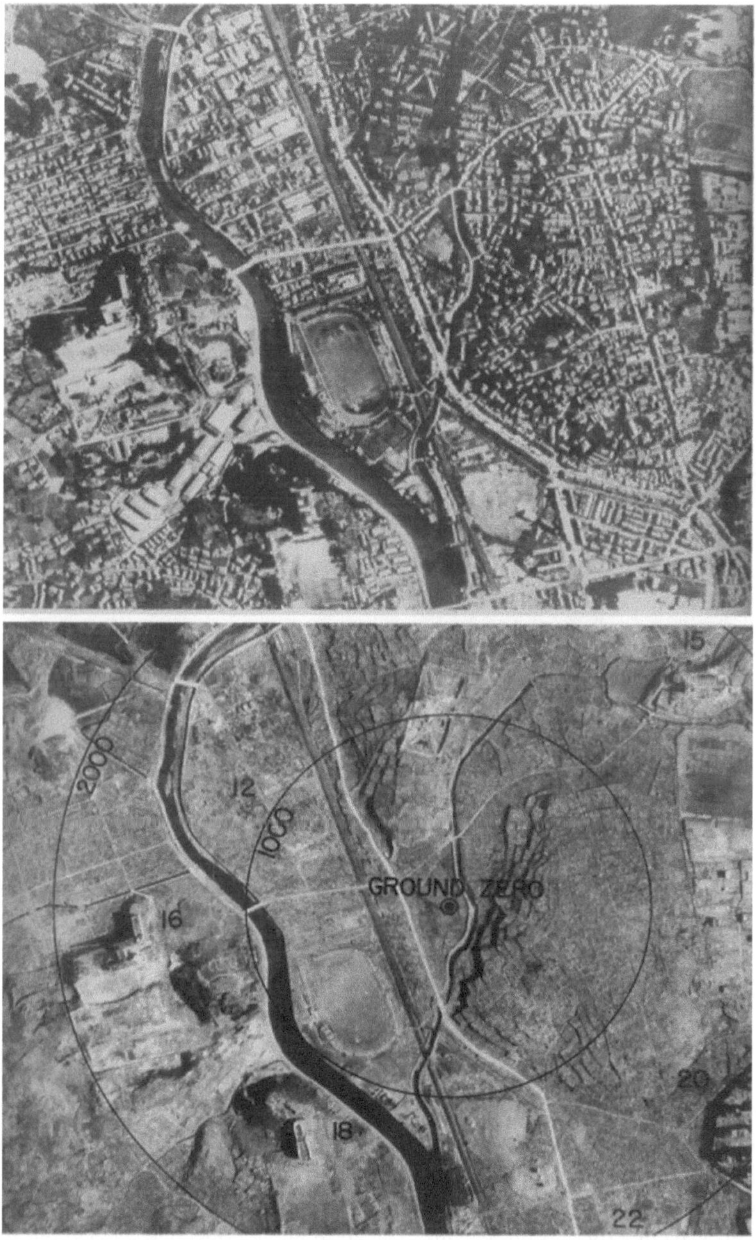

Fig. 8.1 The area of the epicenter 2 days before (*upper*) and 3 days after (*lower*) the atomic bombing. Taken by the US forces

Fig. 8.2 A full-page opinion ad against the nuclear tests in France placed in the local news section of Le Monde; the original photo is on the *left*

testing several times starting in September of the same year. The Japan Federation of Bar Associations and Mayors for Peace issued a statement of protest.

The Nagasaki University School of Medicine, which had experienced catastrophes from the atomic bombing, attempted to wage an intense campaign against the series of nuclear tests by France. As shown in Fig. 8.2, a full-page opinion ad against the nuclear tests in France titled, *Can Nuclear Weapons Bring Peace?* was placed in the local news section (page 7) of *Le Monde* on January 28, 1996. The statement was accompanied by a photo titled, *Two brothers in search of their parents,* taken by Yosuke Yamahata (1917–1968) about 2.2 km from Nagasaki Station on August 10, 1945, to depict a protest by the preceptors of the Nagasaki University School of Medicine against nuclear tests. Initially, those involved contemplated posting the statement from the whole of the Nagasaki University School of Medicine but eventually posted it under the names of voluntary professors, assistant professors, lecturers, and assistants for certain reasons. The sixth nuclear test was carried out the day after this statement appeared in the newspaper, but, subsequently, the nuclear tests ended earlier than expected after French President Chirac announced the end of France's nuclear testing. I think that only the Nagasaki University School of Medicine as an atomic bomb survivor could grasp the devastating damage at that time and the multiyear and considerable impacts brought by the atomic bomb, and I think that we were able to make a much stronger appeal to the French government based on the facts. I still regret that our efforts ended up as a protest only by volunteers in consideration of political nuances and international influences. No record was found about this activity in the memorial bulletin of the 150th anniversary of the foundation of the Nagasaki University School of Medicine published in March 2009.

It took more than 11 years from the time of the atomic bombing until support for the atomic bomb survivors in Hiroshima and Nagasaki begun, and during that period the crew members of a Japanese tuna fishing boat, the *Daigo Fukuryu Maru*, were also exposed to radiation by the nuclear fallout near a US nuclear test site. Meanwhile, the central government had defied citizens' relentless efforts and demands for action in Nagasaki and Hiroshima. First in March 1957, about eleven and half years after the atomic bombing, the Act for Atomic Bomb Survivors' Medical Treatment [*Genshi Bakudan Hibakusya no Iryo-to ni kansuru Horitu*] was enacted and came into force on April 1 of the same year [repealed on July 1, 1995, under the newly enacted Atomic Bomb Survivors' Support Law (*Genshi Bakudan Hibakusya ni taisuru Engo ni kansuru Horitu*)]. Through several hardships thereafter, people gradually started to consider the gravity of the psychological impacts as well as the physical impacts on the atomic bomb survivors. Table 8.1 summarizes the major processes of establishing designation criteria for *atomic bomb survivors* based on support for the psychological impacts.

Table 8.1 Chronology: 1945–2009 events related to Nagasaki atomic bomb survivors

Year	Events
1945–1950	
1945.8.9	11:02 a.m., an atomic bomb dropped on Nagasaki City
1945.9	A survey was held by the Special Committee for Investigation of Atomic Bomb Damages within the Scientific Research Council of Japan (presented first report 30, November)
1945.11	A group of Kyushu University conducted *Neuropsychiatric Research in Atomic Bomb Survivors in Nagasaki*
1948.8	The Atomic Bomb Casualty Commission (ABCC) initiated surveys in Nagasaki and the Nagasaki branch laboratory of the National Institute of Health was established in the site
1949.5	The Atomic Bomb Museum opened at the epicenter in the Nagasaki Peace Park
1951–1960	
1954.3.1	The USA conducted a H-bomb test at Bikini Atoll, Marshall Islands. Japanese tuna fishing vessel *Fukuryu Maru No. 5* was exposed to radiation
1955.4	The Nagasaki International Cultural Hall opened and the atomic bomb museum was relocated in the building
1956	A group of Nagasaki University conducted *Psychiatric Research in Atomic Bomb Survivors*
1957.4.1	*The Act for Atomic Bomb Survivors Medical Treatment* was promulgated and went into effect (revised in 1960)
1958.5.28	Nagasaki Atomic Bomb Hospital opened. (Hiroshima atomic bomb Hospital opened in 1956. Both were transferred to Japanese Red Cross Society in April, 1969)

(continued)

Table 8.1 (continued)

Year	Events
1960.7	The Nagasaki Research Society for Late-effects of Atomic Bombs was established
1960.10	Hiroshima and Nagasaki (Prefectures and Cities) carried out an atomic bomb survivors' survey
1961–1970	
1962.4	The Atomic Disease Institute was established in Nagasaki University School of Medicine
1965.11.1	The Ministry of Health and Welfare undertook the Japanese government's first nationwide survey of actual conditions of atomic bomb victims
1968.5.20	The Atomic Bomb Survivors Special Measures Law was promulgated
1971–1980	
1972.11	Nagasaki Prefecture/City conducted *Basic Survey of Atomic Bomb Survivors and Families*
1975.4	The ABCC was reorganized to inaugurate the Radiation Effects Research Foundation
1975.9	The Ministry of Health and Welfare undertook the second nationwide survey of actual conditions of atomic bomb victims
1979.6	The Study Group on Basic Problems of Atomic Bomb Survivors (Kihonkon) or an advisory panel to the Minister of Health and Welfare was setup
1981–1990	
1980.12.11	The Study Group on Basic Problems of Atomic Bomb Survivors (Kihonkon) offered a written statement
1982.2	The Ministry of Health and Welfare held the survey on atomic bomb victims. (The one tenth of the atomic bomb victims in Hiroshima City and Nagasaki City was included)
1985.10	The Ministry of Health and Welfare conducted the third nationwide survey of actual conditions of atomic bomb victims
1987.10	Nagasaki City conducted *Survey of Elderly Atomic Bomb Survivors* (*survivors aged 60 and over living alone and aged 70 and over living only with spouse*)
1991–2000	
1991.6	*A Report on Residual Radioactivity Derived from the Plutonium Atomic Bomb Dropped on Nagasaki* was submitted
1992.9	Nagasaki City conducted "Survey on Health and Livelihood of Atomic Bomb Survivors"
1993	*A group of Nagasaki University initiated* "Interview Research for Improvement of Mental Health in Atomic Bomb Survivors"
1994.4.	A group of Nagasaki University initiated a scientific research project of the Ministry of Education, Culture, Sports, Science and Technology, a study on *Prevalence of Mental Disorders in A-bomb victims of 50 years ago – as a pilot study for field research*
1994.12.9	The Atomic Bomb Survivors' Support Law passed The Atomic Bomb Survivors' Support Law went into effect
1995.7.1	The Ministry of Health and Welfare conducted the fourth nationwide survey of actual conditions of atomic bomb victims
1995.11	Nagasaki City and neighboring 6 towns conducted *Testimony Study on Undesignated Exposed Areas*
1999.7	A report of the Testimony Study on Undesignated Exposed Areas was published Nagasaki City hosted and held the Nagasaki Atomic Attack Symposium in Tokyo

(continued)

Table 8.1 (continued)

Year	Events
2000.4	An Investigative Commission for Local Testimony Study Reports on Undesignated Exposed Areas was setup in the Ministry of Health and Welfare
2000.7	Nagasaki City hosted and held the Nagasaki Atomic Attack Symposium in Tokyo
2000.10	An Investigative Commission for Local Testimony Study Reports on Undesignated Exposed Areas was setup in the Ministry of Health and Welfare
2001–2010	
2001.3.12–3.30	The T. IKikkawa group of the said Commission conducted a field survey of *the Study on Health Damage Assessment in Association with PTSD*, etc.
2001.8.3	The said Commission submitted the final report
2001.12.18	The Ministry of Health and Welfare published the policies for atomic bomb survivors based on the above report
2002.4.1	The Atomic Bomb Survivors' Support Law and ministerial ordinances relevant to people who experienced bombing were amended and the implementation guidelines were established to roll out the projects
2003.9.1	Nagasaki Prefecture/City established the Commission for Surveys on Actual Conditions of People who Experienced the Bombing
2004.1.-2.	A Field Survey on Actual Conditions of People who Experienced the Bombing (living in areas in a 12-km radius from the epicenter) was held
2004.3.	The Commission for Surveys on Actual Conditions of People who Experienced the Bombing submitted the survey report to the Ministry of Health and Welfare
2004.4.	Nagasaki Prefecture/City submitted the petition for *Abolishment of the Residential Requirements* to the Central Government
2004.10.	The *Investigative Commission for Studies on Mental Damages to People who Experienced the Bombing* was set
2004.12.10	The said Investigative Commission submitted the final opinion
2004.12.27	Nagasaki Prefecture/City set up the Investigative Commission for Expanded Designation of the Exposed Areas (special medical examination areas)
2005.6.1	The project of the Studies on Mental Damages to People who Experienced the Bombing started
2005.11	The Ministry of Health and Welfare conducted the fifth nationwide survey of actual conditions of atomic bomb victims
2007.8–2008.3	The aggregate analysis of reexamined people who experienced the bombing was conducted
2008.5	The report and petition of the Investigative Commission for Expanded Designation of the Exposed Areas (special medical examination areas) were submitted to the Ministry of Health and Welfare
2009.4	*Memory of the bombing experience* was excluded from the criteria for psychiatric care beneficiaries

Atomic bomb survivors (*hibakusha*) are defined in the Law Pertaining to the Support of Atomic Bomb Survivors [Atomic Bomb Survivors' Support Law (or The Atomic Bomb Survivors' Assistance Act), effective July 1, 1995] to include *those who were directly exposed*, *those who moved in the exposed areas*, *those who aided the injured and disposed of the dead*, and *fetuses* as shown in Table 8.2. Table 8.3 shows trends in the number of the *hibakusha* in Nagasaki.

Table 8.2 Definition of atomic bomb survivors (Hibakusha, Article 1 of the Atomic Bomb Survivors' Support Law)

Category	Definition	Number of Hibakusha in Nagasaki (as of March 31, 1974)
1	Those who were exposed directly (within the former city area and surrounding town and villages)	60,521
2	Those who entered the city within approximately 2 km of the hypocenter within 2 weeks of the explosion	21,003
3	Those who were engaged in disposal of the dead or in relief work	722
4	Children of survivors certified in categories 1–3 above who were in utero at the time of the bombing	1,043
Total		83,289

Table 8.3 Trends in the number of the Hibakusya, in Nagasaki City

	Category				
Year	1	2	3	4	Total
1957					66,882
1960					70,390
1965					76,620
1970					82,344
1975	59,443	21,290	945	1,027	82,705
1980	53,254	21,041	3,501	977	78,773
1985	55,376	12,172	3,725	1,229	72,502
1990	50,181	11,306	3,726	1,235	66,448
1995	44,909	10,112	3,688	1,141	59,850
2000	39,053	8,695	3,504	1,071	52,323
2005	35,871	8,013	3,453	1,146	48,483
2010	30,336	6,385	3,057	1,130	40,908

Those who were directly exposed, as defined in the Act for Atomic Bomb Sufferers' Medical Treatment enacted in April 1957, are limited to people who experienced the A-bomb in the then following areas in Nagasaki (atomic *bombsites*, designated in 1952): (1) Nagasaki City, (2) Oura, Koura, Motomura, Koe, Koebaru in Fukuda Village, Nishi-Sonogi-district, and (3) Koda, Yoshimuta in Nagayo Village, Nishi-Sonogi-district. *Those who moved in the exposed areas* include people who entered the area about 2 km from the epicenter (designated as *move-in areas*) within 2 weeks after the bombing (by August 23, 1945 for Nagasaki). *Those who aided the injured and disposed of the dead* include people who were in circumstances in which they were likely to be subject to nuclear radiation at the time of the bombing and thereafter and meet the screening criteria set out in Article 1, item (iii) of the Atomic Bomb Survivors' Support Law (or The Atomic Bomb Survivors' Assistance Act). Lastly, *fetuses* includes unborn babies in the wombs of those who were directly exposed, of those who moved in the exposed areas, and of those who aided the injured and disposed of the dead, born by June 3, 1946, in Nagasaki. The *Atomic Bomb Survivor's Handbook* system was established to identify atomic bomb survivors and to keep track of their health conditions.

After the review of the exposed areas in October 1974 and September 1976, *special medical examination areas* were designated, and the *first-class medical examination certificate* system was established to provide health examinations in these areas. The certificate was switched to the *Atomic Bomb Survivor's Handbook* if those who lived there were afflicted with specific diseases. Figure 8.3 shows the area exposed to atomic bomb radiation.

Using a psychoanalytic method, Robert J. Lifton (a psychiatrist in the USA, born in 1926) [2] described extensively the profound conflict in survivors and the process of overcoming their experience that were extracted through his interviews with the survivors in Hiroshima City conducted in 1962, underscoring how intense the survivors' mental impacts were.

His records also comprise some information obtained from the survivors in Nagasaki. Survivors' experiences with conflicts were also expressed in many literary works, though details are omitted here.

The first psychiatric research on survivors in Nagasaki was conducted by a group from Kyushu University (N. Okumura and H. Hikida) [3], not by staff of Nagasaki

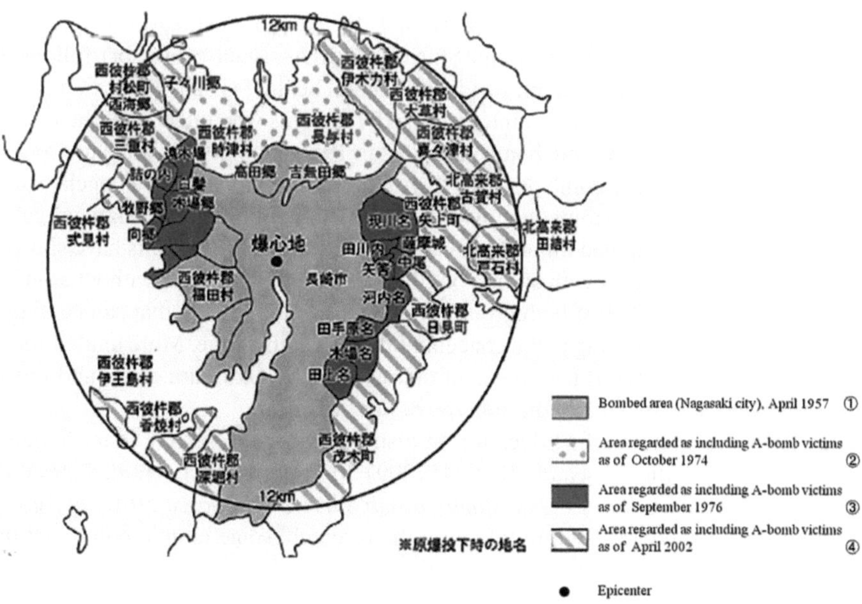

Fig. 8.3 A map of the area exposed to atomic bomb radiation in Nagasaki

University. The research was conducted in 50 patients randomly selected from 192 patients who were in the then Omura National Hospital 3 months after the atomic bombing and investigated the progression of neuropsychiatric symptoms based on their clinical charts. The majority of the subjects presented with *atomic bomb disease* symptoms as a radiological consequence immediately after the bombing and during the earlier phase. Those patients were assumed to be presenting with significant pathological states of psychogenic reaction due to an extremely catastrophic experience, but only a few were identified in such a state, because most of the people with such an experience had probably passed away. The research reported that precise data on situation and frequency could not be obtained. After 1 month, their physical status gradually became stable, and, subsequently, classical neurasthenic symptoms were increasingly observed, which subsided somewhat after 2 months and thereafter. However, most of their complaints remained unsolved.

Physical and social loss experience from the bombing caused long-term and psychological stresses. For example, such stresses may be accelerated by the vulnerable basis of the sufferers' living conditions and fatal psychosocial impacts, including physical factors such as anxiety about keloid scars and being at risk for malignant diseases, distress about microcephaly patients exposed in the mother's womb and their family members as well as fear of its inheritability, lowered social abilities due to acquired disorders, economic difficulties due to increased medical expenses, loss of family members due to death, damaged and destroyed houses, issues of orphans, isolation, despair, and feelings of guilt feeling after survival, and prejudice and discrimination by nonatomic-bomb victims. Despite their attempt at physical and life recovery, they might have been overwhelmed by the multiple damages to daily life as well as the psychological losses.

A group from Nagasaki University led by Taneo Nishikawa (Professor, Department of Neuropsychiatry) and Shiro Tsuiki (Assistant professor) followed the research of Okumura et al. on survivors' mental health [4]. Their research compared frequencies of neurotic symptoms in 7,287 patients who underwent the medical examination for the atomic bomb survivors in 1956 and were assigned to groups with and without atomic bomb diseases (involving symptoms such as alopecia, subcutaneous hemorrhage, stomatitis, pyrexia, diarrhea, vomiting, anorexia, and general fatigue) that developed immediately after the bombing. The results have shown that 9.7 % of the subjects with atomic bomb disease, 3.9 % of those without atomic bomb disease, and 7.3 % of both groups combined were in a state that can be diagnosed as *neurosis* according to the concept of disease at that time. More unidentified complaints were observed than those in the research by Okumura et al., and most neurosis was categorized into the subtype of *neurasthenia*.

In 1995, we obtained the Education Ministry's Science Research Grant (Grant-in-Aid for Scientific Research B, 07457209) for our research entitled *Mental Disorder Prevalence Rates in the Atomic Bomb Survivors Exposed 50 Years Ago – as a pilot field study* [5]. Before initiating the research, some atomic bomb victim groups had published results of various field studies conducted by nonexperts on the survivors' mental health in Nagasaki in addition to the studies by Okumura et al. and Nishikawa et al. I had a strong desire at that time to establish more scientific and epidemiological knowledge and conduct a field study differentiated from those by nonexperts.

In 1993, a pilot study entitled *Interview Survey for Improved Mental Health of Atomic Bomb Survivors* [6] was initiated to assess the feasibility of the scientific research supported by the Ministry of Education, Science, Sports, and Culture. This pilot study demonstrated the viability of studying the correlation of the distance from the epicenter and GHQ-12 scores in 80 survivors before moving into the preparation phase for the main study. We initiated case-finding from the spring of 1994 with the cooperation of relevant internal and external staff of the Nagasaki University School of Medicine, including Professor Masao Tomonaga (Department of Hematology, Molecular Medicine Unit, Atomic Bomb Disease Institute, Nagasaki University), Assistant Professor Mariko Mine (Division of Scientific Data Registry Atomic Bomb Disease Institute Nagasaki University), Assistant Sumihisa Honda (Division of Scientific Data Registry Atomic Bomb Disease Institute), and Dr. Masuko Tagawa from the Health Management Center Nagasaki Atomic Bomb Council.

As of March 1995, there were 59,470 atomic bomb survivors living in Nagasaki (22,765 men and 36,705 women), and distances from the epicenter were known for 58,980 (22,525 men and 36,455 women) of them. The subjects included in this research were 7,670 survivors and second-generation of survivors, combined, living in Nagasaki (3,216 men and 4,454 women) who received medical examinations at the Nagasaki Health Management Center for Atomic Bomb Survivors between October 1994 and August 1996 and consented to be included in this research. The distance from the epicenter was able to be determined for 6,042 of them. Table 8.4 shows the distribution of the survivors and the included subjects in this first step study by distance. It was not statistically significant, but more survivors who were exposed nearer the epicenter were included in the first step study.

The subjects for the secondary and tertiary surveys were selected from the subjects included in the first step study based on the results from the first study with GHQ-12, which was used in the aforementioned PPGHC study. Subjects selected for the secondary survey underwent the Composite International Diagnostic Interview (CIDI) developed by WHO, and those selected for the tertiary survey underwent open-ended interviews conducted by psychiatrists. Out of 445 subjects selected for the secondary survey, the consents were obtained from 226 subjects (106 men and 120 women) and the distances from the epicenter were 173 subjects (77 men and 96 women). A total of 212 subjects (100 men and 112 women) were included in the tertiary survey. The following results were obtained from those subjects in 1997.

Table 8.5 shows the distribution of the first step study results according to the GHQ-12 scoring scale. Comparison with PPGHC data (see Table 7.6) shows that there are more low-scoring groups, and the frequency of high-scoring groups is low, suggesting that the subjects had milder symptoms than internal medicine outpatients.

Comparison by distance from the epicenter, shown in Table 8.6, revealed that the closer the distance to the epicenter, the more deteriorated physical and mental health condition the subjects had, suggesting an astonishing result of the continuing relationship between the distance from the epicenter and overall health level.

Table 8.4 Distribution by distance of the survivors and the included subjects in the first step study [6]

	Up to 2.0 km		2.1–3.0 km		3.1 km and up		Subtotal		Total	
	Survivor	Subject	Survivor	Subject	Survivor	Subject	Survivor	Subject	Survivor	Subject
Male	2,812	423 (15.0)	3,469	470 (13.5)	16,244	1,457 (9.0)	22,525	2,350 (10.4)	22,765	3,216 (14.1)
Female	4,449	609 (13.7)	6,405	815 (12.7)	25,601	2,268 (8.9)	36,455	3,692 (10.1)	36,705	4,454 (12.1)
Total	7,261	1,032 (14.2)	9,874	1,285 (13.0)	41,845	3,725 (8.9)	58,980	6,042 (10.2)	59,470	7,670 (12.9)

Table 8.5 Distribution of GHQ-12 scores in the first study subjects who underwent the medical examination for the atomic bomb survivors

	0–1	2–3	4–	Total
Male	2,589 (80.5)	338 (10.5)	289 (9.0)	3,216
Female	3,502 (78.6)	529 (11.9)	423 (9.5)	4,454
All	6,091 (79.4)	867 (11.3)	712 (9.3)	7,670

Table 8.6 Distribution of GHQ-12 scores in the first study subjects who underwent the medical examination for the atomic bomb survivors

Distance from the epicenter	2.0 km or less	2.1–3.0 km	3.1 km and more
Mean score			
Male	1.123 ± 2.196	0.987 ± 1.832	0.960 ± 1.880
Female	1.123 ± 1.971	1.026 ± 1.960	0.925 ± 1.803
All	1.123 ± 2.065	1.012 ± 1.913	0.939 ± 1.837
The rate of high score subjects (scoring more than 4 out of 12)	9.35 %	9.05 %	8.05 %

Table 8.7 Frequencies of symptoms by CIDI diagnosis in the secondary study subjects

CIDI/ICD-10F code diagnoses	Male	Female	All
F32/33 Depression	3	3	6
F34 Persistent mood disorders	2	3	4
F38.10 Other recurrent mood disorders	1	2	4
F40.2 Specific phobias	7	0	1
F41.1 Generalized anxiety disorder	3	14	21
F41.2 Mixed anxiety and depressive disorder	5	3	6
F45.0/.1/.3/.8/.9 Somatoform disorders	5	12	17
F45.2 Hypochondriacal disorder	1	2	7
Total number of cases	27	39	66

From the secondary step survey, the psychiatric disorders listed in Table 8.7 (limited to common mental diseases) were identified. In the tertiary step survey, 77 subjects (38 men and 39 women) were diagnosed with any psychiatric disorders in the open-ended interview by psychiatrists, as shown in Table 8.8. However, one male subject diagnosed with mental retardation (F70.0, mild mental retardation) of deniable association with atomic bomb exposure was excluded, and a total of 76 patients had psychiatric disorder.

These results and data from the tertiary study showed that the estimated prevalence rate of psychiatric disorders in all the subjects (7,670) was 11.6 and 19.6 %, respectively (see Table 8.9). Further discussion is necessary to determine whether these prevalence rate of the subjects is higher or lower than that in the general population, because no comparable data are available in Japan; however, the study could at least identify cases with a variety of physical and mental impacts even after more

Table 8.8 Results of the tertiary study (clinical diagnosis) in the atomic bomb survivors

	Sex		Age				Distance			GHQ		
ICD-10F	M	F	–59	60–	70–	80–	–2.0	–3.0	3.1–	L	M	H
F01.9/F03 Vascular dementia	1	1	1	0	0	1	0	0	1	0	0	2
F07.0 Organic personality disorder	0	1	1	0	0	0	0	0	1	0	1	0
F10.2 Dependence syndrome due to use of alcohol	2	0	2	0	0	0	0	0	2	0	1	1
F10.5 Psychotic disorder due to use of alcohol	3	0	1	2	0	0	0	1	0	0	0	3
F20 Schizophrenia	1	1	1	1	0	0	0	1	1	1	0	1
F22 Persistent delusional disorders	1	0	1	0	0	0	0	1	0	0	0	1
F32.0/.1/.2 Depressive episode	7	8	7	3	4	1	2	4	6	0	2	13
F33.0/.1/.2/.4 Recurrent depressive disorder	0	7	1	5	1	0	0	2	4	0	0	7
F34.0 Cyclothymia	0	1	1	0	0	0	0	0	1	0	0	1
F34.1 Dysthymia	4	2	2	4	0	0	1	0	3	0	1	5
F41.0 Panic disorder	3	0	3	0	0	0	0	0	2	1	0	2
F41.1 Generalized anxiety disorder	2	2	3	1	0	0	1	0	1	0	1	3
F41.2 Mixed anxiety and depressive disorder	0	1	0	0	1	0	0	0	1	0	0	1
F41.9 Anxiety disorder, unspecified	1	0	1	0	0	0	0	0	1	0	0	1
F43.2 Adjustment disorders	6	7	6	3	3	1	5	0	6	0	2	11
F44.0 Dissociative amnesia	0	1	0	1	0	0	1	0	0	0	0	1
F45.0 Somatization disorder	0	1	0	1	0	0	1	0	0	0	0	1

F45.2 Hypochondriacal disorder	3	1	1	1	2	0	0	0	2	0	0	4
F45.3 Somatoform autonomic dysfunction	1	3	3	1	0	0	0	1	1	0	0	4
F45.4 Persistent somatoform pain disorder	1	0	0	1	0	0	1	0	0	0	0	1
F51.0/.8/.9 Nonorganic sleep disorders	4	4	1	4	3	0	3	0	3	2	3	3
F54 Psychological and behavioral factors associated with disorders or diseases classified elsewhere	1	0	0	0	1	0	0	0	1	0	0	1
F70.0 Mild mental retardation	1	0	0	0	1	0	0	0	1	0	1	0
Total	42	41	36	28	16	3	15	10	38	4	12	67
Total[a]	38	39	32	27	15	3	14	10	35	3	12	62

Age=50–59, 60–69, 70–79, 80–; distance=–2.0, 2.1–3.0, 3.1–; GHQ=low 0–1, middle 2–3, high 4–

The following cases had at least two diagnoses:

F01.9+41.0; F 55–4, F10.2+41.0; M 49 4.0 12, F32.0+41.1; M 48 4.0 8

F32.2+45.2; M 72–6, F41.0+51.0; M 56 4.5 1, F44.0+45.0; F 64 1.1 9

[a]Overlapping cases are combined into each one case

Table 8.9 Estimated lifetime prevalence rate (%) suggested by CIDI diagnosis and clinical diagnosis in the survivors who received the medical examination

GHQ-12	CIDI diagnosis	Clinical diagnosis
Low score arm (0–1)	6.94	11.11
Middle score arm (2–3)	27.78	44.44
High score arm (4–)	31.62	61.86
All	11.59	19.59

朝日新聞 1997年(平成9年)8月3日 日曜日 41636号 (日刊)

朝日新聞

被爆者 52年後も心に傷

爆心に近かった人ほど不安感

長崎大調査 精神面ケア重要

第79回 全国高校野球選手権大会

朝日新聞大阪本社

紙面から

Fig. 8.4 Psychological influences on atomic bomb survivors

than 50 years since the atomic bombing. Also, the prevalence rate is considered to be quite high in comparison with the prevalence of any mental problems in internal medicine outpatients observed in the PPGHC study (14.8 %) and the prevalence (monthly 4.1 %, annually 7.5 %, and lifetime 17.0 %) of mental disorder in all the subjects and (2.8 %, 4.4 %, and 10.6 %, respectively) in the subjects over 65 years of age in the WHO/WHM-J study [7].

These results reported in 1997 [8] generated strong media attention and appeared not only in local newspapers but also in national newspapers (Asahi Shimbun), as illustrated in Fig. 8.4. I was also requested to appear on a welfare educational program of NHK for an interview.

Health damage to the survivors, including not only well-recognized physical impacts but also persisting psychological impacts, has become widely known through international conferences.

Based on the results, the increased awareness of our Nagasaki University research group was extended to the health damage to those who lived in the surrounding area (the concentric circle area in a 12-km radius from the epicenter, in Fig. 8.3; ④ area), of the areas designated as exposed (*12-km north-south length and 7-km east-west length from the epicente*r in Fig. 8.3; the ①, ②, and ③ areas designated as exposed in the Atomic Bomb Survivors' Support Law). Professor Yasuyuki Ohta et al. released *Listen to our Internal Wounds ! – Report on the testimony from the undesignated exposed areas* [9] in 2000 in cooperation with the support office of Nagasaki City for people who experienced atomic bombing in the expanded areas.

The report introduced the earnest appeals of those who were living in the undesignated exposed areas with psychiatric distress and symptoms of posttraumatic stress disorder (PTSD) described as evidence-based medicine (EBM) in *Results of Analysis on Testified Psychological Impacts* [10] by Ohta. The report indicated that a survey on the health conditions of the inhabitants, damage suffered at bombing, testimonies on damage, and the semi-structured interview on PTSD in 7,082 inhabitants of the undesignated exposed areas showed a 6.4 % lifetime prevalence rate of PSTD (PSTD incompletely satisfying the scale was 18.3 %). No other field research in the general population was available at that time. But this prevalence rate appears to be quite high in comparison with US reports on prevalence rates in the general population, such as 1.3 % by Davidson et al. [11] and 2.6 % by Shore et al. [12]. The same methodology was not used in the studies on the survivors and on the inhabitants of the undesignated exposed areas, but these results indicated that the mental health problems in the inhabitants of the undesignated exposed areas could not be ignored.

A collaboration of the prefectural and municipal governments of Nagasaki with researchers decided to make a strong call on the central government for equivalent support for the inhabitants of undesignated areas to that for the survivors. Before this happened, the prefectural and municipal governments of Nagasaki had repeatedly requested actions against local distortion.

However, the Study Group on Basic Problems of Atomic Bomb Survivors (commonly known as *Kihonkon*; Chairman, Professor Seiji Kaya), published in 1980 [13], delivered its view on governmental action for distortion correction, saying that *Expansion of exposed areas with the intention to maintain a balance between the old and new designated areas may cause an unexpected feeling of inequality in the involved, resulting in an unnecessary enlargement,..., unless the expansion process is based on scientific and rational evidence.* The following statements saying: *Considering that atomic bomb survivors have suffered a 'particular casualty' that has a specificity in most of its essentials and an extent to be clearly differentiated from the other war victims, the central government has to take measures being applicable and appropriate to the actual status of damages from the standpoint of governmental compensation in a broad sense,* explaining the governmental compensation in a broad sense as *adequate compensation* equivalent to the degree of damage each survivor has, regardless of the illegality, criminal intent, or negligence of causative behaviors, describes the outcome accountability for health damages from radiation that is *particular casualty*. The Diet resolution (Lower House Committee on Health and Welfare) accompanying the enactment of the Atomic Bomb Survivors' Support Law was adopted in 1994 and specified: *In designating exposed areas, the scientific nature and rationality should be taken into account based on the results of the research results on health hazards from radiation.* The prefectural and municipal governments of Nagasaki made an appeal to the central government for the expansion of designated areas based on the study report and other supportive data on residual plutonium, but the appeal was rejected as not satisfying the *scientific and rational level required by Kihonkon.*

We took the opportunity to report our detailed study results on physical and psychological health damages on the atomic bomb survivors and inhabitants in surrounding areas in the Nagasaki Atomic Attack Symposium hosted by Nagasaki City and held in Tokyo in July 2000. In August of the same year, the Prime Minister at that time indicated his intention to have our testimony and study reports scrutinized in his response to a petition from Nagasaki after the Peace Memorial Ceremony, and the Minister of Health and Welfare brought some measures to Nagasaki in the same month.

Subsequently, an Investigative Commission for Local Testimony Study Reports on Undesignated Exposed Areas (consisting of Chairman Professor Wataru Mori and 11 panel members, including Professor Emeritus Shigenobu Nagataki at Nagasaki University and myself, selected from Nagasaki) was established in October under the supervision of the chief of the Healthcare Bureau in the Ministry of Health and Welfare.

During the deliberation by the above investigative commission, a comparison of part of the testimony report—(*1*) *whether the disability identified is specifically caused by radiation*, (*2*) *whether the investigational methods used are based on satisfactorily scientific and rational evidence, and* (*3*) *whether the study results are acceptable for the general nation* to the relevant description in the publication of *Kihonkon*—led to the conclusion of the need for an additional study. Accordingly, a group for research on the mental health status associated with the distress related to the atomic bomb explosion was organized as a special scientific research study of the Ministry of Health and Welfare, and a panel member of the commission, Takehiko Kikkawa, Director Emeritus of the National Institute of Mental Health, National Center of Neurology and Psychiatry, was selected as the principal researcher.

The study conducted by this group was completely independent from the one conducted in Nagasaki and spanned about one month from March 2001. The results submitted by the study group to the investigative commission have demonstrated the scientific robustness of the study [14]. The final commission meeting was held in August of the same year and suggested, based on the results, that the survivors group living in the relevant area might have trauma caused by the A-bomb experience and persisting as anxiety, which might affect their mental health and lead to their reduced physical health. Their low health level was determined to have been highly possibly caused by the anxiety associated with their A-bomb experience, not by the direct radiation exposed at bombing. In December of the same year, the Minister of Health and Welfare made a statement to *designate the undesignated exposed areas in a 12-km radius from the epicenter as special health examination areas* and to cover the medical expenses for relevant diseases in accordance with the Atomic Bomb Survivors' Support Law. The Ministry also manifested support for a new category designated as *people who experienced bombing*. On April 1, 2002, the whole project involving the enforcement of the revised Atomic Bomb Survivors' Support Law, as well as ministerial ordinances, implementation of medical examination as the seventeenth article of the Atomic Bomb Survivors' Support Law stipulates, establishment of guidelines for medical expenses and counseling on diseases caused by psychological impacts through A-bomb experience, was rolled out.

This project, launched with a focus on mental health, was not wholly accepted by local citizens. They most frequently claimed that they were now included in the sufferers according to the new designation criteria but were still distinguished from being an *atomic bomb survivor*. I myself who have been involved in the past process cannot refute but have to accept the claim; on the other hand, however, I have to realize how difficult it is to establish new evidence that explains the effects associated with radiation exposure. Another major claim was that they feel discomfort in seeking consultation with a psychiatrist and being diagnosed with mental disorders. It reflects their distorted view (prejudice) on psychiatric treatment that they cannot dare to contact a psychiatrist and to be diagnosed with any mental disorders in order to be eligible for medical expense support. Such a distorted view is likely to gradually diminish once they can understand our assertion on the psychosomatic correlation that mental health damages involve physical health damages.

After the launch of this project, more citizens than expected took advantage of this opportunity, while some abuses were also observed. Concurrently, inequality in this project came to the surface; the inclusion criteria limited inhabitants to those currently living in areas within a 12-km radius from the epicenters as eligible inhabitants and excluded those who moved from the above areas. In order to eliminate this obstacle, the prefectural and municipal governments of Nagasaki repeatedly requested the central government to *abolish the residential requirements*. In response to the Health and Welfare Minister's statement against the request that *new scientific conclusions are prerequisite to address this issue*, a local *investigative commission for expanded designation of the exposed areas (special medical examination areas) was set up accordingly, and a Field Survey on Actual Conditions of People Who Experienced the Bombing (living in areas in a 12-km radius from the epicenter)* was conducted in 2004 by local researchers belonging to the subcommittee as the members who experienced the bombing. This survey used the method employed in the aforementioned study of the Kikkawa group and included subjects living in various areas in Nagasaki Prefecture and meeting the criteria for people who experienced the bombing.

Coinciding with the submission of the survey results to the Ministry of Health, Labor, and Welfare, the secondary *Investigative Commission for Studies on Mental Damages to People who experienced the bombing (Chairman; Yasuhito Sasaki, Nobuyuki Takahashi; Professor of Nagasaki International University, and me as panel member from Nagasaki)* [15] was established under the supervision of the chief of the Healthcare Bureau on October 6, 2004. The panel members from Nagasaki faced the direct fire of critical attacks for abuses in business operations and other issues during five commissions held over about 2 months. As a result, the commission formed a conclusive opinion on the abolishment of residential requirements and the increased responsibility of psychiatrists in clinical practice.

Following the course of the first project, the secondary project started on June 1, 2005, with more intense focus on medical treatment (particularly psychiatric treatment) than *health care*, aiming at reaching an earlier treatment effect. After the review, those meeting criteria for people who experienced the bombing but are suffering a memory loss were excluded by the Ministry of Health, Labor, and Welfare, raising new concerns among local citizens.

In a local project commission in Nagasaki, prefectural and municipal governments of Nagasaki established evidence for the abolishment of the requirement for *memory* through scrutiny of the past study and survey results, ultimately convincing the central government to agree to it.

In my lengthy description on how mental health support reached the survivors and sufferers in Nagasaki, introduction of essentially complex backgrounds was bypassed and the results obtained were only presented in brief. The support system in Nagasaki is not considered complete, and various issues still remain. And I as a citizen of Nagasaki had been involved in studies focusing on the survivors' support system in Nagasaki, but there are presumably other issues also in Hiroshima, expecting a solution developed by the central, prefectural, and municipal governments together in which opinions and expertise of researchers are incorporated.

In conjunction with atomic bomb survivors (hibakusha), there are also other themes that arouse my interest, such as mental health damages to sufferers of nuclear accidents occurring at nuclear power plants. Nagasaki University has been continuing studies on radiation risk control in its own research facilities for disorder and disability treatment after the A-bombing, inviting and accepting researchers from Chernobyl. Some collaborative studies were implemented with psychiatrists from that site.

References

1. The Commission for the Conservation of Atomic Bomb. Report on the post-disaster situations. In: Atomic Bomb Survivors' Affairs Dept. Research Section 2009 Overview for the Project of Atomic-Bomb Survivor's Affair'. Nagasaki: Nagasaki City Hall; 1950 (in Japanese).
2. Lifton RJ. Death in life: survivors of Hiroshima, New York. New York: Random House Inc; 1967 [republished 1991 by the University of North Carolina Press, Chapel Hill (translated into Japanese by Masui T, Yuasa N, Ochi M et al, (2009) Asahi Newspaper, Tokyo, 1971, Iwanami, Tokyo)].
3. Okumura N, Hikida H. Results of psycho-neurological studies on atomic bomb survivors. Kyusyu Neuropsychiatry. 1949;1:50–2 (in Japanese).
4. Nishikawa T, Tsuiki S. Psychiatric investigations of atomic bomb survivors. Nagasaki Med J. 1961;36:717–22 (in Japanese).

5. Nakane Y, Tomonaga M, Mine M, Takada K, Honda S. Prevalence rates of mental disorders among survivors exposed atomic bombing 50 years ago – a pilot study for a community survey. 1995 Research Grants for Scientific Research Ministry of Education General Basic B; Results Report; 1997 (in Japanese).
6. Nakane Y, Honda S, Tomonaga M. Interview survey on mental health improvement to atomic bomb survivors and its analysis. In: 1996 Atomic Bomb Diseases Study Group (Chief; Shigematsu I). Ministry of Health Sciences Research Grants, 1997 Report of the Research; 1997. p. 19–22 (in Japanese).
7. Nakane Y. Epidemiological study of mental disorders among aged people. Jpn J Geriatr Psychiatry. 2009;20(Special issue III):14–9 (in Japanese).
8. Nakane Y, Tomonaga M. Mental disorders among atomic bomb survivors and their relationship to A-bomb exposure. In: 1996 Atomic Bomb Diseases Study Group (Chief; Shigematsu I). Ministry of Health Sciences Research Grants, 1997 Report of the Research; 1998. p. 33–7 (in Japanese).
9. Atomic Bomb Survivors' Affairs Department Research Section, Nagasaki City Hall. Listen to our mental suffering. 1999 report of testimony survey in the undesignated area of radiation exposure (designated areas for medical examination). Nagasaki: Atomic Bomb Survivors' Affairs Dept, Research Section, Nagasaki City Hall; 2000. p. 1–141 (in Japanese).
10. Ohta Y. Results of analysis on testified psychological impacts. In: Atomic Bomb Survivor's Affairs Dept., Research Section, Listen to our mental suffering – 1999 report of the testimony survey in the undesignated area of radiation exposure (designated areas for medical examination). Nagasaki: Nagasaki City Hall; 2000. p. 87–95 (in Japanese).
11. Davidson JR, Hughes D, Blazer DG, George LK. Post-traumatic stress disorder in the community: an epidemiological study. Psychol Med. 1991;21:713–21.
12. Shore JH, Vollmer WM, Tatum EL. Community patterns of posttraumatic stress disorders. J Nerv Ment Dis. 1989;177:681–5.
13. Council of Basic Issues for Hibakusha Affairs ("*Genbaku-Hibakusya-Taisaku-kihonmondai-Kondankai; KIHONKON*" Chairman: Kaya S). Report of the study group on basic problems of atomic bomb survivors. 1980. http://shoruisouko.xsrv.jp/kntk/ss3.pdf#search. Accessed April 2012 (in Japanese).
14. Kim Y, Tsutsumi, Izutsu T, Kawamura N, Miyazaki T, Kikkawa T. Persistent distress after psychological exposure to the Nagasaki atomic bomb explosion. Br J Psychiatry. 2011;199:411–6.
15. Committee on Reform of the Research Projects of Mental Impact due to Atomic Bomb Experiences ("*HIBAKU-TAIKENSYA-Seishin-Eikyou-toh-Chousa-Kenkyu-Jigyo-no-Arikata-ni-kansuru-Kentokai*" Chairman: Sasaki Y). Report of the investigative commission for studies on mental damages to people who experienced the bombing. 2004. http://www.mhlw.go.jp/shingi/2004/10/s1006-2.html. Accessed April 2012 (in Japanese).

Epilogue

Further Developments in Social Psychiatry and Psychiatric Epidemiology

Looking back on my research history so far, I have been interested in a variety of issues and have conducted various research studies. However, I think that ultimately a coherent philosophy of research was evident. When a study is conducted at a laboratory in a university, I consider the ideal situation to be the one where the researcher gives priority to pursuing themes that fit within the stated goal of his or her laboratory for contributing to the public, and then to carry out his or her own personal research interests.

Thus, the public research themes for me were social psychiatry studies and cross-national collaborative studies of various mental disorders, mainly schizophrenia and mood disorders, which the World Health Organization has set as the priority for research, and also independent biological studies concerning mental disorders, which the Department of Neuropsychiatry at Nagasaki University collectively conducted. On the other hand, my personal research themes were the study of diseases on the border between neurology and psychiatry, such as epilepsy, and the study of psychiatric and behavioral disorders in children and adolescents, based on my experience at the children's consultation center where I worked part-time for a long time. I spent a substantial amount of time on each of these themes, and I assume that I reasonably achieved them.

In this book, I have described the details of social psychiatry, especially the psychiatric epidemiology of schizophrenia and depression and primary care-related psychiatric care. Epidemiological study of genetics in the neurological field, which I was greatly interested in during my younger days, is almost impossible to undertake, and, in addition, relevant cases are seldom seen in daily clinical practice, because neurology has now become completely independent from psychiatry.

I served as a psychiatrist in the Department of Neuropsychiatry at Nagasaki University School of Medicine for about 40 years and have set my direction as the study of social psychiatry in clinical psychiatric practice. As I was in charge of the

Y. Nakane, *Progress in Social Psychiatry in Japan: An Approach to Psychiatric Epidemiology*, DOI 10.1007/978-4-431-54103-5, © Springer Japan 2012

courses in this area, I have been involved in teaching students in the medical school and other related departments, and training and supervising postgraduate students and young psychiatrists as well. In this situation, there are several themes which aren't possible to discuss in this book.

One of the most important of such themes is forensic psychiatric practice, including psychiatric examination. I kept to my principle of accepting offers to psychiatrically evaluate people who caused serious criminal cases in Nagasaki and nearby cities. I published some of these psychiatric examinations in papers for not only my interest but also with the desire to enlighten other professionals involved in relevant areas. I have written a long article in my own name in the newspaper, which was widely published, concerning the abundance of simplistic views about mental disorders and their incidence, as well as the misuse of terms relating to specific psychiatric disorders and psychiatry without understanding the reality. I wrote it expecting to dispel prejudice against psychiatry and psychiatric care, standing with my senior and colleagues in the university on their antinuclear position as well.

Following my retirement from the teaching staff of the School of Medicine, I became a teaching staff at a private university and was involved in education for mental-health-based health-care professionals. Since then, my research themes have largely shifted from psychiatry to just mental health. My main educational activity, in addition to educating students who major in social welfare, is to enhance general public understanding of the current status of and improvement of prejudice and discrimination against mental disorders and people with such disorders.

The largest study was a comparative study of two countries that examined mental health literacy over approximately 4 years, with assistance from the Grants-in-Aid for Scientific Research from the Ministry of Health, Labor, and Welfare. I investigated whether or not prejudice against mental disorders and toward people with the disorders in Japan was substantial compared to that in Australia, using a common questionnaire. The results showed that "social distance" toward mental disorders and people with the disorders were markedly greater in Japan. I am now investigating the backgrounds and problems in the education that students receive.

I hope that I can continue the research and practice in mental disorders a little longer. The World Health Organization has been promoting the theme of "No development without health; no health without mental health" for several years. I am also greatly supporting this theme, as I think this phrase grasps the important points.

While the importance of evidence-based medicine (EBM) has spread, its value has not been fully established yet in Japan. I think that people who view themselves as researchers should publish evidence through their studies.

However, when I look at the course of study that I set for this book, I am really ashamed, because I realize that, in fact, I myself have not established findings that can fully respond to EBM. I have truly tried to carry out research with the intention of showing various outcomes through the data. Nevertheless, as I knew that there were many data to analyze in more detail, I realized that my research had not reached a stage that can provide conclusive results through finer analysis. It can be said that I have confronted again the difficulty of establishing EBM only after reviewing my own research process.

Nevertheless, the findings reported in this book were obtained from results of not only my research but also collaborative research with many of my colleagues. It cannot be denied that my participation in a few of the studies was quite small. However, in all cases, I would like to express my sincere appreciation to all my colleagues for helping me to complete this book. Moreover, I would like to deeply thank the president and the staff of the publisher, Nakayama Shoten Co., Ltd.

Appendix

Medical Office (*Ikyoku*)

A medical office, or *ikyoku* in Japanese, is generally considered to be an office or lounge for doctors in a hospital. However, an *ikyoku* is a group of doctors which serves important functions beyond such a physical space, functions essential for both academic and medical practice, including training and instruction to younger doctors. The *ikyoku* concept and its structure probably developed from Japan's unique background, which values collective decision-making within a hierarchically organized society. Although originally the *ikyoku* was a private organization in a hospital, it is now commonly recognized as a public body in many university hospitals. Therefore, the head of the *ikyoku* who represents the *ikyoku* is an executive-level doctor in the hospital, and executive meetings consisting of the *ikyoku* heads are held.

A clinical department (chair system) consists of a professor, associate professor, assistant professors, research associates, staff members, clinical trainees (doctors-in-training), and graduate students, and in many university hospitals the *ikyoku* also consists of the same members. In some university hospitals, a professor and associate professor of the clinical department are not included in the constituent members of the *ikyoku*. Where there are a professor and associate professor, decisions against opinions and requests raised by the *ikyoku* are made by the clinical department. Some large general hospitals also have similar *ikyoku* as a collective entity of doctors. These days, the head of the *ikyoku*, who has typically been designated by the professor of the department, is often chosen in an election, recommended by the *ikyoku* members, and approved by the professor.

An individual entering medical school after senior high school and then completing 6 years of education needs to graduate from the medical school and take the National Medical Practitioners Examination. After passing the examination, the individual needs to receive clinical resident training designated by the Ministry of Health, Labor and Welfare in accordance with the law in hospitals for 2 years or more to become a full-fledged doctor. The medical director of a hospital where a

Y. Nakane, *Progress in Social Psychiatry in Japan: An Approach to Psychiatric Epidemiology*, DOI 10.1007/978-4-431-54103-5, © Springer Japan 2012

clinical trainee receives training gives him or her provisional tenure until he or she decides which clinical department to go. Many clinical trainees decide the clinical department they want to go around one year after starting the training. When he/she can receive approval for acceptance by that department, he or she is admitted to the *ikyoku* of that clinical department.

For a clinical trainee, enrollment in a clinical department of his or her choice is, so to speak, the first step to starting his or her career as doctor. He or she becomes a staff member immediately after completing the training or after postgraduate studies at a graduate school of medicine, and will advance to research associate, assistant professor, and associate professor. Enrolled younger doctors receive instructions from seniors in the *ikyoku* step by step in an apprenticeship manner. The *ikyoku* takes care of private and public matters of enrolled doctors.

The significant concerns of *ikyoku* include training younger doctors to develop experts as well as management of human resources. A doctor enrolled in an *ikyoku* can accumulate experience with the assistance of their seniors in a hospital affiliated with the *ikyoku*, while at the same time, the affiliated hospital can secure doctors from the *ikyoku*. The *ikyoku* system seems advantageous for doctors and hospitals, but failure to establish continued harmonious relations between doctors and hospitals may cause troubles. In fact, it has become difficult to maintain cooperative relationship between *ikyoku* and affiliated hospitals due to recent revisions in the training program for younger doctors. Increasing negative opinions on the aforementioned *ikyoku* system may contribute to changing the circumstances more than ever.

Bibliography

Kendell RE, Cooper JE, Gourley AJ, et al. Diagnostic criteria of American and British psychiatrists. Arch Gen Psychiatry. 1971;25:123–30.

Lieberman RP. Psychosocial management of schizophrenia overcoming disability and handicap. Jpn J Soc Psychiatry (Shakai Seishin Igaku). 1988;11:269–72.

Nakane Y. Conditions of children with Down's syndrome in Nagasaki. Bull Nagasaki Child Guidance Center (Nagasaki Chuo Jido Sodansho Kiyo). 1977;1:45–9 (the text of a particular title is only in Japanese).

Nakane Y. Down's syndrome in Nagasaki – the shift of childbirth to younger mothers. J Child Health (Shoni Hoken Kenkyu). 1978;37:39–43 (the text of a particular title is only in Japanese).

Nakane Y. Epidemiology in psychiatry. In: Kakeda K, Shimazono Y, Okuma T, Takahashi R, Hozaki H, editors. Handbook of modern psychiatry. Supplement for '87-A. Tokyo: Nakayama-shoten; 1987. p. 31–57 (the text of a particular title is only in Japanese).

Nakane Y. Methodology of epidemiological research. Jpn J Clin Psychiatry (Rinsho Seishin-Igaku). 1999;28:39–48 (the text of a particular title is only in Japanese).

Nakane Y. Promotion of research on psychiatric epidemiology. Jpn Bull Soc Psychiatry. 2001;9:269–86 (the text of a particular title is only in Japanese).

Nakane Y. Globalization and diversity – challenges of social psychiatry. SCJ FORUM (Gakujutu no Doko, Science Council of Japan). 2005a;10:58–60 (the text of a particular title is only in Japanese).

Nakane Y. A history of social psychiatry. In: Japanese Society of Social Psychiatry, editor. Social psychiatry. Tokyo: Igaku-Shoin; 2009a. p. 2–15 (the text of a particular title is only in Japanese).

Nakane Y. Preface: toward social psychiatry from community psychiatry. Jpn Bull Soc Psychiatry. 2009b;18:176–7 (the text of a particular title is only in Japanese).

Ohara K, Okagami K, Yoshimatsu K, Nakane Y, Oda S. Talking about social psychiatry. Jpn Bull Soc Psychiatry (Nihon Shakaiseishin-igakkai Zasshi). 1993;1:3–16 (the text of a particular title is only in Japanese).

Fujii I, Hirota N, Yuzuriha T. Clinical research of depressives treated in an outpatient clinic of a general medical setting. Bulletin of Research Funded by Daiwa Health Foundation Daiwa Herusu Zaidan Josei Kenkyu Gyoseki-shu. 1981;5:189–93 (the text of a particular title is only in Japanese).

Giggs JA, Cooper JE. Ecological structure and the distribution of schizophrenia and affective psychoses in Nottingham. Br J Psychiatry. 1987;151:627–33.

Honjo K, Kawakami N, Takeshima T, et al. Social class inequalities in self-rated health and their gender and age group difference. Jpn J Epidemiol. 2006;16:223–32.

Y. Nakane, *Progress in Social Psychiatry in Japan: An Approach to Psychiatric Epidemiology*, DOI 10.1007/978-4-431-54103-5, © Springer Japan 2012

Ichinose H, Nakane Y, Nakane H, et al. Relationship between ultralong-term outcome (after 28 years) and duration of untreated psychosis. Acta Med Nagasaki. 2009;54:59–66.

Kaneko S, Battino D, Andermann E, et al. Congenital malformations due to antiepileptic drugs. Epilepsy Res. 1999;33:145–58.

Kawakami N, Takeshima T, Ono Y, et al. Twelve-month prevalence, severity, and treatment of common mental disorders in communities in Japan: preliminary findings from the World Mental Health Japan Survey 2002–2003. Psychiatry Clin Neurosci. 2005;59:441–52.

Kinoshita H, Nakane Y, Nakane H, et al. Schizophrenia study: influence of the duration of untreated psychosis on long-term outcome. Acta Med Nagasaki. 2005;50:17–22.

Naganuma Y, Tachimori H, Kawakami N, et al. Twelve-month use of mental health services in Japan: findings from the World Mental Health Japan Survey 2002–2003. Psychiatry Clin Neurosci. 2006;60:240–8.

Nakane Y. Congenital malformation among infants of epiletic mothers treated during pregnancy – the report of a collaborative study group in Japan. Folia Psychiatr Neurol Jpn. 1979;33:363–9.

Nakane Y. Clinical course and prognosis of schizophrenia. Clin Psychiatry. 2003;45:575–82 (the text of a particular title is only in Japanese).

Nakane Y. Long-term outcome (WHO collaborative studies). Schizophrenia Frontier. 2005b;6:281–8 (the text of a particular title is only in Japanese).

Nakane Y. Life events relating to onset, relapse, and clinical course. Psychiatr Neurol Jpn. 2007;109:751–8 (the text of a particular title is only in Japanese).

Nakane Y. Experiences in psychiatric epidemiology research – at the time of the WHO collaborative study on the determinants of outcome of severe mental disorders. Kyusyu Neuropsychiatry. 2009c;55:162–9 (the text of a particular title is only in Japanese).

Nakane Y, Ohta Y, Okazaki Y, Michitsuji S, Takada K. A ten-year long-term follow-up study of patients with first episode of schizophrenia: outcome and socio-psycho-biological factors. 1992 Research Grants for the Scientific Research Ministry of Education, General B, Results Report 1993 (the text of a particular title is only in Japanese).

Nakane Y, Ohta Y, Radford MHB. Epidemiological studies of schizophrenia in Japan. Schizophrenia Bull. 1992b;18:75–84.

Nakane Y, Takahashi R, Tominaga Y, et al. Study of the rate of schizophrenia in Nagasaki city. Clin Psychiatry. 1985;27:771–81 (the text of a particular title is only in Japanese).

Ohta Y, Araki K, Nakane Y, Takahashi R. Criteria for schizophrenic disorder in DSM-III and the prodromal symptoms. Clin Psychiatry. 1984;26:589–98 (the text of a particular title is only in Japanese).

Ohta Y, Nakane Y. Diagnostic problems of schizophrenia according to symptomatological assessment by family members and psychiatrists: on first contact with schizophrenics with acute onset and ones with insidious onset as subjects. Jpn J Soc Psychiatry. 1987;10:60–7 (the text of a particular title is only in Japanese).

Ohta Y, Nakane Y. Urban ecology and incidence rate of schizophrenia results from a study in Nagasaki city. Jpn Bull Soc Psychiatry. 1993;1:31–52 (the text of a particular title is only in Japanese).

Ohta Y, Nakane Y, Nishihara J, Takemoto T. Ecological structure and incidence rates of schizophrenia in Nagasaki city. Acta Psychiatr Scand. 1992;86:113–20.

Okuma T, Takahashi R, Nakane Y, et al. Teratogenic effects of antiepileptic drugs: a collaborative study in Japan. Adv Neurol Sci. 1979;23:1247–63 (the text of a particular title is only in Japanese).

Ohta Y, Tominaga Y, Araki K, Michitsuji S, Nakane Y. Study on the relationship between patients' family members' recognition of schizophrenia and two-year outcome. Jpn J Soc Psychiatry. 1986;9:57–65 (the text of a particular title is only in Japanese).

Ohta Y, Tominaga Y, Nakane Y. Relationship between family members' recognition of early abnormalities and motivation for seeking psychiatric services by the family. Jpn J Soc Psychiatry. 1985;8:232–8 (the text of a particular title is only in Japanese).

Ohta Y, Tominaga Y, Nakane Y. A study on the relationship between family members' affective attitude toward schizophrenic patients and interpretation of etiology and two year-outcome. Jpn J Clin Psychiatry. 1987;16:1703–9 (the text of a particular title is only in Japanese).

Sartorius N, Jablensky A, Korten A, et al. Early manifestations and first-contact incidence of schizophrenia in different cultures. Psychol Med. 1986;16:909–28.

Sata M, Yoshitake K, Utsunomiya H, Hatada K, Ohtsuka T, Sugasaki H, Nakane Y. Factors affecting disability in patients attending the internal medicine departments of general hospitals. Psychiatry Clin Neurosci. 1999;53:611–20.

Utsunomiya H, Yoshitake K, Ohtsuka T, Sugasaki H, Sata M, Hatada K, Nakane Y. Methodological issues of an epidemiological study from the results of a study on psychological problems seen in outpatients visiting the internal medicine departments of general hospitals. Psychiatry Clin Neurosci. 1999;53:25–32.

Wing JK, Cooper JE, Sartorius N. Measurement and classification of psychiatric symptoms – an instruction manual for the PSE and CATEGO Program. Cambridge: Cambridge University Press; 1974 (translated into Japanese by R. Takahashi, Y. Nakane et al. (1981) Igaku-Shoin, Tokyo).

World Health Organization. The international pilot study of schizophrenia. Geneva: WHO; 1973.

Yoshitake K, Utsunomiya H, Sonoda Y, et al. Health perceptions of outpatients visiting the internal medicine department of general hospitals. Jpn Bull Soc Psychiatry. 1998;7:153–63 (the text of a particular title is only in Japanese).

Mental Health Research for A-Bomb Survivors (Hibakusha)

Honda S, Mine M, Imamura Y, Yoshitake K, Nakane Y, Tagawa M, Tomonaga M. Mental health survey of Nagasaki atomic bomb survivors. Nagasaki Med J. 1996;71(special edition):322–4 (the text of a particular title is only in Japanese).

Honda S, Mine M, Tomonaga M, et al. Mental health survey of Nagasaki atomic-bomb survivors – relationship between current mental health status and exposed condition at the time of bomb ing. Nagasaki Med J. 1998;73(special edition):237–9 (the text of a particular title is only in Japanese).

Honda S, Mine M, Tomonaga M, Nakane Y, Tagawa M, Shibata Y. Accuracy of memory in a survey of acute symptoms of atomic bombing. J Hiroshima Med Assoc (Hiroshima Igaku). 2002;55:187–9 (the text of a particular title is only in Japanese).

Honda S, Shibata Y, Mine M, Imamura Y, Tagawa M, Nakane Y, Tomonaga M. Mental health conditions among atomic bomb survivors in Nagasaki. Psychiatry Clin Neurosci. 2002b;56:575–83.

Ichinose H, Nakane H, Kinoshita H, et al. Regarding the mental health of people who experience the atomic bomb residing in Nagasaki. Nagasaki Med J. 2006;81(special edition):49–52 (the text of a particular title is only in Japanese).

Mine M, Honda S, Hata T, et al. Mental health for elderly A-bomb survivors. Nagasaki Med J. 1994;69(special edition):313–8 (the text of a particular title is only in Japanese).

Nakane Y. Disaster psychiatry. In: Know the Psychiatry. AERA MOOK 15. Tokyo: Asahi Shinbun; 1996. p. 68–9 (the text of a particular title is only in Japanese).

Nakane Y. Welfare report in the Japanese archipelago: care of mind for A-bomb survivors. NHK Seminar on Social Welfare '98. 1998;4–7:98–101 (Reported on NHK Educational TV, December 18, 1997) (the text of a particular title is only in Japanese).

Nakane Y. Mental health among atomic bomb survivors-considering the consequences of the atomic bombing. Papers of the Lecture for Nagasaki Symposium on Atomic Bombing, Atomic Bomb Survivors' Affairs Department, Research Section. Nagasaki: Nagasaki City Hall; 2000. p. 7–13 (the text of a particular title is only in Japanese).

Nakane Y. Psychiatric approach and psychological care for atomic bomb survivors. J Hiroshima Med Assoc. 2002;55:133–7 (the text of a particular title is only in Japanese).
Nakane Y, Hatada K, Honda S, et al.. The influence of atomic bomb disaster on mental health. Nagasaki Med J. 1996;71(special edition):161–71 (the text of a particular title is only in Japanese).
Nakane Y. The process to be authorized with psychological disorder in victims of the atomic bomb. Clin Psychiatry. 2006a;48:273–86 (the text of a particular title is only in Japanese).
Nakane Y, Honda S, Mine M, Tomonaga M, Tagawa M, Imamura Y. The mental health of atomic bomb survivors. In: Nagataki S, Yamashita S, editors. Nagasaki symposium radiation and human health: proposal from Nagasaki. Amsterdam: Elsevier; 1996b. p. 239–49.
Nakane Y, Honda S, Tomonaga M. Interview survey on mental health improvement of atomic bomb survivors and its analysis. In: 1994 Atomic Bomb Diseases Study Group (Chief; Shigematsu I), Ministry of Health Sciences Research Grants, 1995 Report of the Research; 1995. p. 21–3 (the text of a particular title is only in Japanese).
Nakane Y, Honda S, Tomonaga M. Interview survey on mental health improvement of atomic bomb survivors and its analysis. In: 1995 Atomic Bomb Diseases Study Group (Chief; Shigematsu I), Ministry of Health Sciences Research Grants, 1996 Report of the Research; 1996. p. 20–3 (the text of a particular title is only in Japanese).
Nakane Y, Imamura Y, Honda S. Mental health of atomic bomb survivors. Jpn J Psychiatr Treat (Seishinka Chiryougaku). 1998;3:987–92 (the text of a particular title is only in Japanese).
Nakane Y, Imamura Y, Honda S, Yoshitake K. Mental Health of Atomic Bomb Survivors and the University. In: Center for Lifelong Learning Steering Committee for Regional Arts and the University of Nagasaki, Nagasaki Public Lectures Series 11: Regional Affairs and the University. The University of Nagasaki Printing Bureau, Japanese Finance Ministry; 1999. p. 201–8 (the text of a particular title is only in Japanese).
Nakane Y, Ohta Y. The example of linkage with a cancer register. In: Horn T, Giel R, Gulbinat W, Henderson JH, editors. Psychiatric case registers in public health – a worldwide inventory 1960–1985. Amsterdam: Elsevier; 1986. p. 240–5.
Nakane Y, Takada K, Sonoda Y, Yoshitake K, Sugasaki H, Imamura Y. Mental Health Guide for A-Bomb Survivors. Nagasaki Prefecture, Nagasaki: Atomic Bomb Survivors Relief Division, Department of Health and Welfare; 1999 (the text of a particular title is only in Japanese).
The Committee for the Compilation of Materials on Damage Caused by the Atomic Bombs in Hiroshima and Nagasaki (ed.) (1981) Hiroshima and Nagasaki, The Physical, Medical, and Social Effects of the Atomic Bombings, Iwanami Shoten Publishers, Basic Books (Ed. by Hiroshima·Nagasaki City Genbaku-Saigai-shi Henshu-Iinkai (1979) Hiroshima·Nagasaki no Genbaku-saigai, Iwanami-shoten, Tokyo).
Tsuiki S, Ueno K, Segawa K, Kaburagi S. Experiences of the Department of Neuropsychiatry in Nagasaki Medical College for five years after the war. Psychiatr Neurol Jpn. 1951;53:229 (abstract) (the text of a particular title is only in Japanese).

Mental Health in the Nuclear Power Plant Accident

Ikeuchi M, Yamawaki N, Nakane Y, et al. Post-traumatic stress disorder in the Toukai-mura neutron line bombing accident. Nagasaki Med J. 2002;77(special edition):280–3 (the text of a particular title is only in Japanese).
Korol N, Shibata Y, Nakane Y. Psychosomatic health status of children exposed to the Chernobyl accident. Acta Med Nagasaki. 1998;43:57–61.
Kryzharovskaya L, Nakane Y. Mental health of liquidators of the Chernobyl disaster. Acta Med Nagasaki. 1996;41:15–9.

Psychiatric Problems Among Evacuees of Volcanic Eruption

Araki K, Ohta Y, Kawasaki N, Nagaoka K, Nakane Y. Disaster psychiatry (therapy) and disaster mental health (care): methods of support for the survivors. Psychiatr Neurol Jpn. 1996a;98:757–60 (the text of a particular title is only in Japanese).

Araki K, Ohta Y, Kawasaki N, Nagaoka K, Nakane Y. Disaster stress and psychological care – starting at the Unzen-Fugendake eruption. Tokyo: Ishiyaku-Syuppan Ltd.; 1996b (the text of a particular title is only in Japanese).

Araki K, Takahashi R, Nakane Y, Ohta Y, Ishizawa M, Tominaga Y, Uchino J. Natural disaster and mental disorders – psychiatric study on Nagasaki floods (1982). Psychiatr Neurol Jpn. 1985;87:285–302 (the text of a particular title is only in Japanese).

Fujita C, Ohta Y, Nakane Y, Takada K. Investigation of the psychological effects of volcanic eruptions of Mount. Unzen and the establishment of a mental health system. 1993 Research Grants for the Scientific Research Ministry of Education General Basic B; Results Report; 1994 (the text of a particular title is only in Japanese).

Nakane Y. Evidence and practice in social psychiatry. Jpn Bull Soc Psychiatry. 2006b;14(Suppl): 10–22.

Nakane Y. Trends in psychiatry – thinking about the article "Natural disaster and mental disorders." Psychiatr Neurol Jpn. 2007;109:633–6 (the text of a particular title is only in Japanese).

Nakane Y, Aikawa K. Disaster and mental health for children – catastrophic stress and psychiatric problems. Jpn J Children's Adolesc Psychiatry (Jido Seinen Seishin Igaku to Sono Kinsetsu Ryoiki). 1995;36:397–402 (the text of a particular title is only in Japanese).

Nakane Y, Aikawa K. Psychological stress among children exposed to the eruption of the Unzen-Fugen-dake volcano. Jpn J Psychother (Seishin Ryoho). 1996;22:41–50 (the text of a particular title is only in Japanese).

Ohta Y, Araki K, Kawasaki N, Nakane Y, Honda S, Mine M. Long-term mental health measures to evacuated residents: support activities for the experience of the Mount. Unzen eruption. Psychiatr Neurol Jpn. 1995;97:1124–30 (the text of a particular title is only in Japanese).

Ohta Y, Araki K, Kawasaki N, Nakane Y, Mine M, Honda S. Study of psychiatric problems of evacuees from the volcanic eruption of Mount Unzen – factor analysis of the General Health Questionnaire (GHQ-30). Jpn Bull Soc Psychiatry. 1995;3:109–29 (the text of a particular title is only in Japanese).

Ohta Y, Araki K, Kawasaki N, Nakane Y, Honda S, Mine M. Psychological trauma and longitudinal course of psychiatric problems among evacuees of a volcanic eruption. Psychiatry Clin Neurosci. 1998a;52(Suppl):s41–8.

Ohta Y, Araki K, Kawasaki N, Nakane Y, Takemoto T, Honda S, Mine M. Two-year follow-up study of psychiatric problems among evacuees of the Mount Unzen volcanic eruption. Jpn Bull Soc Psychiatry. 1998b;6:197–214 (the text of a particular title is only in Japanese).

Ohta Y, Araki K, Kawasaki N, Nakane Y, Honda S, Mine M. Psychological distress among evacuees of a volcanic eruption in Japan: a follow-up study. Psychiatry Clin Neurosci. 2003;57:105–11.

Takemoto T, Nakane Y, Ohta Y, Moji K, Senju H, Nagata K. Human ecological study on health impacts of volcanic eruption, 1996–1997 Research Grants for Scientific Research Ministry of Education General Basic B-2; Results Report; 1998 (the text of a particular title is only in Japanese).

Takemoto T, Nakane Y, Ohta Y, Senju H, Izumi T, Aoyagi K. Health problems in communities changed by a volcanic disaster, 1998–2001 Research Grants for Scientific Research Ministry of Education General Basic B-2; Results Report; 2002 (the text of a particular title is only in Japanese).

Index

MIX
Papier aus verantwortungsvollen Quellen
Paper from responsible sources
FSC® C105338

If you have any concerns about our products,
you can contact us on
ProductSafety@springernature.com

In case Publisher is established outside the EU,
the EU authorized representative is:
Springer Nature Customer Service Center GmbH
Europaplatz 3, 69115 Heidelberg, Germany

Printed by Libri Plureos GmbH
in Hamburg, Germany